UNDERGROUND WORLDS

UNDERGROUND WORLDS

A Guide to Spectacular Subterranean Places

DAVID FARLEY

BLACK DOG
& LEVENTHAL
PUBLISHERS

Black Dog & Leventhal Publishers
Hachette Book Group
1290 Avenue of the Americas
New York, NY 10104

www.hachettebookgroup.com
www.blackdogandleventhal.com

First Edition: May 2018

Black Dog & Leventhal Publishers is an imprint of Hachette Books, a division of Hachette Book Group. The Black Dog & Leventhal Publishers name and logo are trademarks of Hachette Book Group, Inc.

The publisher is not responsible for websites (or their content) that are not owned by the publisher.

The Hachette Speakers Bureau provides a wide range of authors for speaking events. To find out more, go to www.HachetteSpeakersBureau.com or call (866) 376-6591.

Print book interior design by Red Herring Design.

LCCN: 2017948544
ISBNs: 978-0-316-51402-6 (hardcover), 978-0-316-51400-2 (ebook)
Printed in Malaysia
IM
10 9 8 7 6 5 4 3 2 1

CONTENTS

Aeneas and the Sibyl in the Underworld, 1598, by Jan Brueghel the Elder (1568–1625).

INTRODUCTION

HUMANS HAVE GONE UNDERGROUND SINCE TIME IMMEMORIAL. Or at least for the last forty-three thousand years or so. That's when archaeologists believe the oldest known mine in human history was created. The Ngwenya mine, also known as the Lion Cavern, is located in Swaziland in southern Africa. It was discovered in 1970 when archaeologists Adrian Boshier and Peter Beaumont, having unearthed some ancient digging tools in the area, began rooting around a mountain and found the mine, which is rich in iron ore.

Humans chipped away at the cavern's interior until about twenty-three thousand years ago, extracting hematite presumably used to create red ochre for art and cosmetics, after which the mine lay dormant, waiting to be rediscovered. To put this into context, consider that the mine was possibly first used in 41,000 BC, sixteen thousand years before humans crossed the Bering Strait from Asia to North America and thirty-one thousand years before the first Agricultural Revolution.

It's possible that we've been digging into the ground since much earlier than that. Or at least sheltering in caves, naturally made "homes" that are conveniently protected from the elements and predators. In 2014, archaeologists announced that after conducting a series of new tests on cave art on the Indonesian island of Sulawesi, they determined the paintings—of handprints and something called a "pig deer"—were much older than previously believed. They dated the paintings to be at least thirty-five thousand years old. But certainly humans took up the troglodyte lifestyle long before it occurred to us to decorate our living room cave walls with images of pig deer.

Besides literally going below the surface of the earth, we have a long, creative history of figuratively burrowing ourselves into the underworld. The subterranean world has long captured our imaginations. Jules Verne's *A Journey to the Center of the Earth,* J. R. R. Tolkien's Middle-earth, and E. M. Forster's increasingly prophetic short story "The Machine Stops," in which earthlings are forced to live underground because the earth has become uninhabitable, are a few examples that come to mind. Perhaps the most famous portrayal is Dante Alighieri's *Inferno*, the first third of his epic *The Divine Comedy*, where the protagonist goes on a journey to the netherworlds—Hell sectioned into nine circles.

In both fact and fiction, human life belowground has been motivated by utilitarian reasons—shelter, safety, and access to minerals or water—but also for more divine reasons: every culture, every religion and faith, has their notion of the underworld. In Buddhism, it's Naraka; in ancient Greece, it was Hades; the Incas called it Ukhu Pacha; in Persian mythology, it's Duzakh; and of course, in Christian lore, it's called Hell. Whether feared as punishment for the sins of the living or revered as the eternal realm of the dead, it's often believed to be a physical, not just metaphysical, place.

In this book, the motivations behind creating subterranean worlds are as diverse as the cultures they come from. In Derinkuyu, located in Cappadocia, Turkey, humans honeycombed the subterrestrial world to hide from invaders. In Coober Pedy, Australia, they holed up to flee the extreme heat. In the Cu Chi tunnels outside of the Vietnamese metropolis that locals still call Saigon, the North Vietnamese soldiers outlasted American troops, David outwitting the technologically superior Goliath via the use of underground passageways. Similarly, the Sarajevo War Tunnel, constructed in the early 1990s during the Bosnian conflict to transport medical supplies, food, and civilians into a part of the city that had been blocked off by Bosnian Serbs, was a desperate (and successful) instance of using the underground for survival.

Whatever has led us to the netherworlds—figuratively and literally, for shelter or survival, to commune with the dead or to seek treasure—the world beneath our feet remains a mystery and a fascination.

And sometimes the reasons remain a mystery, as they perhaps were even to the people who first broke ground. These inspired excavators were often guided by a sense of the divine. Such was the case with Levon Arakelyan in Yerevan, Armenia, who was driven to dig by what he claimed to be the voice of God. Initially, he began digging a potato cellar as a favor for his wife, but as the voice in his head directed him to continue, he just kept chipping into the earth. It's the same with the devotees of the Federation of Damanhur, located outside of Turin, Italy, who collectively saw a falling star and took it as the sign they'd been waiting for that it was time to put shovel to earth. Why? To build a system of temples dedicated to unlocking the hidden potential of the human spirit. In the end, both Arakelyan and the disciples of Damanhur created their own versions of paradise, subterrestrial sanctuaries that started in the mind and became a physical reality.

The Ngwenya mine, or Lion Cavern, in Swaziland.

Digging into the earth is not just a thing of the past. Far from it. Recent digs have given new meaning to the term "groundbreaking." The G-Cans, located on the periphery of Tokyo, is the world's largest sewer system, protecting the Japanese metropolis from severe flooding. It houses cisterns big enough for Godzilla—as well as Rodan, Mothra, and a couple of his other on-screen pals—to comfortably bathe. Perhaps a bit more aesthetically pleasing is the *Dome of Light,* the largest single piece of glasswork on the planet, located in the Formosa Boulevard subway station in Kaohsiung, Taiwan. And no discussion of twenty-first-century subterranean projects would be complete without including the world's first underground park, the Lowline, located below New York City's Lower East Side. The park uses cutting-edge technology to shoot natural light twenty-five feet below the bustling Big Apple streets.

Our contemporary fascination with the underground not only is borne out of military, religious, or engineering necessity but could also be a kind of evolutionary nostalgia, going back to the "caveman" days—that simplified view of hunter-gatherer societies when early humans took to caves to eat their pre–Agricultural Revolution diets (which happen to be currently in vogue) and doodle on the cave walls. Today, we've turned caves into everything from restaurants (Ristorante Grotto Palazzese in southern Italy) to hotels (Corte San Pietro in Matera) to bars (Cave Bar More in Dubrovnik), all for our own recreation and amusement.

Whatever has led us to the netherworlds—figuratively and literally, for shelter or survival, to commune with the dead or to seek treasure—the world beneath our feet remains a mystery and a fascination.

From the
ENGLISH SEA
to the
MEDITERRANEAN

A partial view of the Tomb of the Valerii, one of the Necropolis's most luxurious, which was owned by Valerius Philumenus and Valeria Galatia.

THE VATICAN Necropolis

VATICAN CITY

O N DECEMBER 23, 1950, POPE PIUS XII MADE AN ANNOUNCEMENT that shook the Christian world. On a radio broadcast heard around the globe, he said, "The tomb of the Prince of the Apostles has been found." He paused for dramatic effect and then added, "Such is the final conclusion after all the labor and study of these years."

By "labor and study," he meant a massive undertaking that began eleven years earlier, an excavation as deep as twenty-five feet underneath Saint Peter's Basilica. It had been rumored for centuries that the altar was positioned above the suspected grave of the fisherman of Galilee, Jesus's apostle Peter, also considered to be the first pope. And now the Vatican was intent on confirming it. Led by German monsignor Ludwig Kaas, the excavation unearthed an entire city of the dead. Buried in the fourth century to provide a flat surface for the basilica, the massive necropolis contains Roman tombs, basalt-paved roads, and intricate two-thousand-year-old mosaics. But what mattered most is that they found what they were looking for: the burial place of Saint Peter.

Today, visitors can walk through a side entrance of the basilica, past the colorfully dressed Swiss Guards, and descend to the depths of the Vatican for a guided tour of this necropolis. It's not easy, though. Only 250 people per day are allowed in the Vatican Necropolis, or Scavi, as it's referred to in Italian, which is located five stories beneath the basilica. After applying online—being flexible with dates can increase your chances of being selected—lucky visitors simply receive an e-mail stating what

Situated directly under the dome of the basilica, Saint Peter's Baldachin was commissioned by Pope Urban VIII in 1623.

date and time they should appear. To put it into perspective, consider this: thirty thousand art-loving tourists march through the ornate halls of the nearby Vatican Museums every day.

Vatican Hill, as it was called two millennia ago, is just outside the walls of Rome. It's no accident there's a necropolis on this spot. Romans avoided dead bodies for fear of contamination, so Roman law dictated that all cemeteries must be outside the city walls. Hence the reason the famed catacombs, where early Christian martyrs were buried, lie in the countryside along the Appian Way.

Peter was buried on Vatican Hill because he'd been martyred nearby at the circus of Nero—reputedly crucified upside down—sometime between AD 64 and 67, and Catholic tradition held that the dead be buried as close as possible to where they died. Emperor Constantine decided to build old Saint Peter's, the church that originally stood where Saint Peter's Basilica now resides, on the spot where the apostle was rumored to be buried. The rub, though, was that as the centuries ticked on, no one was certain Constantine was right about the placement of Saint Peter's grave. It's true that in the first few centuries of Christianity, bones—suspected to be those of Saint Peter—were discovered on Vatican Hill. And thus Constantine made his best guess, having part of the cemetery on Vatican Hill filled in with dirt, and the altar of old Saint Peter's positioned on the spot they hoped was Saint Peter's final resting place.

One thousand years after its initial construction, old Saint Peter's, as we know it today, was in poor condition. So the church came up with a plan to build a bigger, better, more extravagant Saint Peter's. After all, it was the late fifteenth century, and the new ideas and grand style of the Renaissance were spreading throughout Europe. Construction took 109 years and saw the reign of twenty different popes. At least ten architects worked on the project, including such illustrious men as Michelangelo and Bramante. For the building, material was stripped from ancient monuments, including tons of marble from the Colosseum and the Forum. In constructing the "new Saint Peter's," which it was called for a few generations, the necropolis beneath was nearly forgotten.

In the late 1870s, workers were preparing the tomb of recently deceased Pope Pius IX in the crypt, where most popes are buried, and found a larger necropolis underneath Saint Peter's. But it wasn't until 1939, when Pope Pius XII ordered a mass excavation to unearth the ruins, that they confirmed what they had hoped all along: Saint Peter's tomb rests below the basilica's altar.

Today tourists roam under Saint Peter's Basilica among the arched and pedimented tombs in various states of disrepair. Some boast incredible black-and-white mosaics

The papal tombs under Saint Peter's Basilica, most of which date from the fifth to the sixteenth centuries.

depicting chariots, warfare, and great moments in Roman history. Most of the graves are pagan. There are altars with legible Latin inscriptions etched into them. And then there are graves with no adornment, most likely the resting spots of slaves.

On June 26, 1968, after researchers tested the bones found within the tomb and determined them to be those of a sixty- to seventy-year-old man, Pope Paul VI announced that the remains were Saint Peter's. But there are many skeptics. After all, in the same tomb, archaeologists also recorded finding the bones of four different individuals and a number of farm animals, leading some experts to question the likelihood that the bones belonged to the saint. The Vatican, though, was convinced.

The tour takes ninety minutes and zigzags down ancient cobbled underground lanes flanked by hundreds of tombs designed with columns and friezes. At the end there's the holy highlight of the tour: from a distance, through a small window, visitors can see inside the famed tomb and view the bones—or so the church hopes—of Saint Peter the Apostle.

In 2003, archaeologists discovered another swath of the Vatican Necropolis, realizing it is much bigger than anyone suspected. Located just next to the Vatican parking garage, the Vatican Necropolis of the Via Triumphalis boasts over one thousand tombs that date from 1 BC to AD 322.

One of the most memorable graves in the Necropolis of the Via Triumphalis is that of Tiberius Natronius Venustus. On it is a bust of a young boy, most likely Tiberius, framed by a small templelike stone structure. The inscription reads that he had lived for four years, four months, and ten days. No cause of death was given. In another place in the necropolis, terra-cotta tubes stick out of the ground. They were used for pouring libations in honor of the dead, a pagan custom that was temporarily adopted and then abandoned by the Christians.

In another place in the necropolis, terra-cotta tubes stick out of the ground. They were used for pouring libations in honor of the dead, a pagan custom that was temporarily adopted and then abandoned by the Christians.

Up above the necropolis, inside Saint Peter's Basilica, the famed Baldacchino di San Pietro, designed by baroque sculptor Gian Lorenzo Bernini, sits directly above the tomb of Saint Peter. The structure, which rises like a canopy almost one hundred feet above the altar, is made from bronze, 90 percent of which was stripped from the Pantheon, the most remarkable architectural relic of Roman antiquity.

The site opened to visitors in 2006, but it's a separate tour and has a separate entrance from the Scavi. Like visitors to the Scavi, you can book a journey into this underworld via the Vatican's website.

DOMUS Aurea

ROME, ITALY

AROUND 1480, A TEENAGE BOY WAS PLAYING ON THE OPPIAN HILL near the Colosseum when he slipped through a crack in the ground and plummeted into the darkness. City authorities, with torches in hand, were lowered into the pit to find the boy and figure out what this mysterious crevasse was hiding. To say they were surprised would be an understatement.

Built by Emperor Nero in the first century, the Domus Aurea rests in the Roman Forum.

As they moved the torches around to shed light on parts of the long-forgotten room, they saw ornately painted frescoes from the Roman era, which were around fifteen hundred years old at the time.

The teenage boy had inadvertently rediscovered the Domus Aurea, or Golden House, the pleasure palace of Roman emperor Nero. After Nero died in AD 68, the space was covered up, and the Baths of Trajan and the Colosseum were built on top of the controversial site.

The emperor built his gargantuan palace on a previously urbanized swath of Rome that had burned down. Over the centuries, some historians have even suggested that Nero himself was clandestinely responsible for the fire, so that he could build on the spot. He hired Severus and Celer, the two best architects in the empire, to design and construct it. He had the city's greatest living artist, Famulus, adorn the walls with frescoes depicting creatures from Greek and Roman mythology, such as the Cyclops. When the palace was finished, Nero publicly dedicated the Domus Aurea by proclaiming, without any trace of irony, "God, I can at last begin to live like a human being."

The palace held a mile-long arcade, and many interior walls were covered in gold and adorned with mother-of-pearl. Hot springs bubbled up into rooms, channeled in from underground streams via fifty-foot aqueducts.

While Nero was living like a human being —finally!—ordinary Romans were sneering as they walked by. For many denizens of the capital, the palace was a symbol of Emperor Nero's luxurious excesses. It became the epicenter of extravagance, the site of posh parties for the city's upper crust. There were three hundred rooms and no sleeping quarters—Nero's own official residence was on the nearby Quirinal Hill.

And then there was just the sight of this prodigious palace. Originally named the House of Passages, it stretched over half a mile from Palatine Hill to Esquiline Hill. But when a portion of the palace was destroyed in a fire, Nero rebuilt the structure and called it Golden House. The courtyard was so large, it was occupied by a 120-foot-tall statue of the emperor himself. The Colosseum actually gets its name from this "colossal" statue and rests on the spot where the palace's famed artificial lake once was. The first-century Roman historian Suetonius wrote that the enormous water basins that made up the lake were "more like a sea than a pool." The palace held a mile-long arcade, and many interior walls were covered in gold and adorned with mother-of-pearl. Hot springs bubbled up into rooms, channeled in from underground streams via fifty-foot aqueducts. The architects also created a domed dining room in which the ceiling would rotate, powered by slaves. While the well-heeled debauched themselves with food and wine, the turning frescoed ceiling would drop rose petals.

Frescoes from one of the rooms in the Domus Aurea, or Golden House.

As the story goes, at one such gathering, so many petals dropped from above that one reveler actually choked to death. Exotic-but-tame beasts roamed the gardens surrounding the palace, and sculptures from Greece and Asia Minor were imported to adorn the grounds.

Amazingly, the lavish palace only stood for four years. After Nero committed suicide in AD 68, his successors tried to expunge him from history. Nero's death marked the end of a turbulent time in Roman history—it ended the Julio-Claudian dynasty, which had commenced with the reign of Augustus and evolved over the years to such levels of extravagance that Romans finally had enough. And so, not long

A view from the underground passageways of the Domus Aurea.

after Nero's demise, the Domus Aurea was dismantled. The lake was drained. The colossal statue of Nero was moved out of sight. The palace was stripped of all its posh ornamentation. And, finally, the vaulted halls and spacious rooms were filled in with dirt, as if the Domus Aurea had never existed. This once highly visible and glorious aboveground structure was condemned to lie beneath the surface of Rome for centuries.

What we now know as the Oppian Hill sat mostly undisturbed and abandoned for centuries, with the Domus Aurea hidden underneath, a silent witness to the dismantling of the Roman Empire. Though Rome's population plummeted from one million at its most powerful zenith to just fifty thousand at the start of the ninth century, the Domus Aurea got a new life when that Renaissance-era teenage boy fell into the crevasse.

A true living palimpsest, Rome is a jewel box of archaeological underground treasures. Each time the city builds a new subway line, it takes longer than planned because of subterranean discoveries. For every ancient site that is uncovered, they must call in archaeologists to carefully dig it out. So in a city dotted with two-thousand-year-old catacombs, why is the Domus Aurea significant?

The unexpected and accidental discovery of the ancient palace and its intact artwork in the late fifteenth century set off a frenzy among central Italy's artists. The original Roman frescoes were shielded from the ravages of passing time, preserving their vivid colors. Soon enough, artists such as Raphael, Michelangelo, Ghirlandaio, and Pinturicchio ventured down to get a look at them, helping to push the Renaissance along. The seeds of the Renaissance had already been planted, but the discovery of the frescoes in the Domus Aurea inspired the craze for classical antiquity. The mosaics are another artistic innovation credited to the Domus Aurea: for the first time, and apparently at Nero's behest, mosaics made it onto the ceilings and the upper part of walls. Before then they were only inlaid in floors. This change would establish a highly influential precedent for Christian art in the following centuries.

Despite the Renaissance-era excitement for the rediscovered Domus Aurea, the remnants of the palace were soon ignored again, but not forgotten. By the eighteenth century, Oppian Hill was covered by a vineyard. And in the 1930s, Mussolini had the hill expanded, creating a park that offered a nice view of the Colosseum.

After a twenty-year restoration, the Domus Aurea opened to the public for the first time in 1999. Six years later, it was shut down, as it was deemed unsafe. Two years later, the doors were flung open once again, but were then closed two years after that when part of the roof crashed down to the floor. Finally, the Italian government got serious and decided to invest about $37.5 million to restore the Domus Aurea. During the renovation, archaeologists unearthed new parts of the palace, including a huge circular dining room. They are still working to restore the three hundred thousand square feet of frescoes—that's more than thirty Sistine Chapels put together.

The Hall of Water in the Temples of Humankind.

FEDERATION *of Damanhur*

TURIN, ITALY

ABOUT THIRTY MILES NORTH OF TURIN, ITALY, IN A SMALL TOWN underneath the towering Alps lies what some people have called the "Eighth Wonder of the World." But standing in front of it, you'd wonder: What's the big deal? That's because the entrance looks more like a back-yard toolshed than anything else.

Welcome to the Federation of Damanhur, a small property in the town of Baldissero Canavese in the Piedmont region, whose residents proclaim it to be an autonomous micronation. The "wonder" in question is a series of underground shrines, called the Temples of Humankind, that the disciples of Damanhur built. The organization, a community of six hundred people, helped fund the project by opening a series of small businesses in town. The temples are an eight-chamber, baroque, five-story subterranean complex one hundred feet below the surface of the earth that looks like the Palace of Versailles and a 1970s hippie commune collided. The high-ceilinged chambers include soaring pillars, colorful floor mosaics, walls bedecked in gold leaf, mirrors, and colorful psychedelic murals. Both Liberace and Tim Leary would have been very comfortable here. The multichamber complex consists of the Blue Temple, the Hall of Water, the Hall of the Earth, the Hall of Spheres, the Hall of Metals, and a massive four-sided pyramid called the Hall of Mirrors, which is covered entirely in—you guessed it—mirrors. Each chamber is more spectacular than the next: the ceiling and upper half of the walls of the Hall of Spheres are covered in twenty-four-karat gold leaf; the Hall of Water, dedicated to the female divine force, is clad in a deep blue and bathed in an aqua-colored, Tiffany-inspired glass cupola ceiling; the Hall of the Earth, made up of two connected circular rooms forming the symbol of infinity, is highlighted by floor-to-ceiling white ceramic columns near the walls and mosaic images of naked men looking triumphant and glorious; in the Hall of Metals, with metal representing every age of humanity, there are pillars bedecked to look like trees. This chamber has four niches in the walls, one of each standing for earth, water, fire, and air.

Founded in 1975 by Oberto Airaudi, a former insurance salesman, the Federation of Damanhur is named after the Egyptian town of Damanhur, located one hundred miles northwest of Cairo, and dedicated to the god Horus. The federation has their own government, currency, schools, newspaper, and tax code. Their belief system draws on pantheistic pagan and Egyptian convictions blended with new age philosophies. They subscribe to the power of human capacity and believe that humans have not fully tapped into their real potential yet, as if we could turn ourselves into superhumans. Maybe this is why they claim to have found a way to time travel. So far, they have not been particularly forthcoming on how they figured it out.

The multichamber complex consists of the Blue Temple, the Hall of Water, the Hall of the Earth, the Hall of Spheres, the Hall of Metals, and a massive four-sided pyramid called the Hall of Mirrors.

The Damanhur disciples have a proclivity toward the paranormal and supernatural. They believe there are rivers of energy, lines flowing around the globe, that humans can tap into to access parts of our brains we have hitherto not reached. Conveniently, their headquarters in the region of Piedmont, Italy, happens to be on the convergence of some of those energy lines—in fact, they say their complex and headquarters in Baldissero Canavese lie on the convergence of four of the eighteen synchronic lines that crisscross the planet. The members adopt Italian names based on flora and fauna: Farfalla (Butterfly), Ananas (Pineapple), Gazza (Magpie), and Macaco, named after a macaque monkey, are the names of just a few members. And when encountering one another on the Damanhur grounds, members great each other not with "hello" or "ciao" but *con te* (or for two or more people, *con voi*), which translates to "I'm with you."

The origins of the Temples of Humankind go back to a warm August night in 1978. Oberto and ten Damanhurians were lounging around a fire. As the official story goes, the group noticed a star falling, leaving behind it a lingering trail of stardust. They had been planning to create an underground temple, and they took this as a sign that it was time to begin. So a few people picked up some shovels and pickaxes and started digging a tunnel into the side of the nearby mountain. In the morning, a few others relieved them. The digging continued until they'd created a massive, cavernous circular space inside the mountain. They built walls, painted some groovy images from floor to ceiling, and laid down a colorful mosaic on the floor and the walls.

Amazingly, this underground temple building was all done in secret. The Italian government did not even know about the project—and thus never granted them permission for it—until 1992, at which time the police demanded to see the temples. But the Damanhurians wouldn't let them inside. After the authorities threatened to blow up the mountain where the complex is located, the devotees finally gave in.

That there's this new age settlement in northern Italy might come as a surprise to some. But scratch the surface of Turin's history, and it's perfectly fitting. Even expected.

The city has long had a reputation for being a center for occult activity. It was rumored, for example, to be home to forty thousand Satanists. There are supposedly five landmarks stretching across the historic city center that make up a pentagram. And if you ask anyone who studies the occult about Turin, they may tell you that the city straddles two of those energy lines that the Damanhurians and new age types believe in. In this case, the axis of white magic (along with Lyon and Prague) and the axis of black magic (which it shares with London and San Francisco). This, if you believe in that sort of thing, would make Turin and its environs one spiritually, energetically powerful place.

It all goes back to the founding and uniting of Italy as a nation in the mid-nineteenth century when Turin became the capital of the new country. In a defiant move against the pope, who was staunchly against unification, Italy's Turin-based king invited persecuted religious groups to come worship in the northern Italian city. Jews and Protestants arrived. Even the Book of Mormon was translated into Italian for the first time in Turin. This cemented the city's reputation as being tolerant of other spiritual ideas. And that never stopped.

In this context, the existence of the Federation of Damanhur near Turin, with its "eighth wonder of the world" underground temples, is not so unusual, after all. If only they'd just share the secret to time travel!

Today visitors can see the temples and the complex where followers still live and work. Every Sunday, they offer tours of four of the eight temples, but only in Italian. For guests who want to stay longer, there are also one-day visits that include a tour of the temples, lectures on the Damanhur brand of spirituality, and guided group-meditation sessions.

The Hall of Earth in the
Temples of Humankind.

CATACOMBS Across EUROPE

ROME, Italy

PARIS, France

PALERMO, Italy

T HE WORD "CATACOMBS" IS possibly derived from Latin for "cata tumbas," or "among the tombs," and it refers to the Appian Way just outside of Rome, where one could—and still can—stroll down the basalt-stone-paved street flanked by tombs of wealthy Roman Christians, the ancient movers and shakers who refused the usual Roman custom of cremation. Today there are many places in the world where one can walk among catacombs built for the purpose of entombing the dead, including Odessa, London, Lima, Melbourne, Alexandria, and Znojmo in the Czech Republic.

But nowhere is more associated with the word "catacombs" than, of course, Rome.

Because most Romans in the two or three centuries after the death of Christ were still pagan, their dead bodies were cremated, as was the custom in the Roman Empire. But with the rise of Christianity and the rejection of cremation (after all, you had to have a body intact for the apocalypse and thus resurrection), Christian bodies were buried deep underground outside the walls of Rome. And so the catacombs were born around the end of the second century.

Many of the bodies of the early Christian martyrs were spirited away to the catacombs. Cults to the martyrs emerged, as early Christians prayed in their underground cemeteries, hoping that the tombs were a channel to God, that the martyr or saint would receive the request and whisper it into the ear of divinity.

The two most popular catacombs today—and the largest—are out on the Appian Way, the ancient Roman road that stretched four hundred miles from Rome to Brindisi. Visiting the catacombs of either San Sebastiano or San Callisto should provide a solid primer on the history of early Christianity in Rome and how the newish cult of Christ ended up becoming an official religion in the capital of the empire. In San Callisto, for example, the catacombs run four layers deep and twelve miles long, holding the tombs and graves of five hundred thousand people. Many of the bones have been taken from the catacombs, as a medieval craze for relics inspired greedy tomb raiders hoping to sell the bones of early martyrs to monasteries or churches, who would pay a heavenly sum. One of the most popular underground lairs in San Callisto is called the "little Vatican." Still lit today by oil-burning lamps, it held the remains of nine popes who led the clandestine church until the third century. On the interior walls of the catacombs, once-esoteric secret messages have been etched: an anchor was disguised as a cross, the fish was meant to symbolize Jesus, and a bird was meant to be a phoenix, which represented the resurrection of Christ.

But the catacombs in Rome are not just limited to the Appian Way. In the last millennia, dozens of catacombs

One of many grave chambers in the Roman catacombs of San Sebastiano.

have been rediscovered and dug up around the city. No one really knows how many yet-to-be-found underground cemeteries are sprinkled around Rome. For now, though, the Via Appia Antica catacombs, and its miles of underground tomb-flanked tunnels, should be enough to satisfy anyone with a desire to commune with the dead.

PERHAPS SECOND ONLY TO Rome's necropolis, Paris's underground necropolis is well-known around the world. Today it's a popular spot for both tourists and locals. One of the most famous tales of the Parisian underworld goes back to the French Revolution:

Enter Philibert Aspairt. On November 3, 1793, this career doorman at the Val-de-Grâce hospital in Paris decided to descend into the catacombs from an entryway in the hospital. No one knows why. And no one knows what happened to Philibert. Except that eleven years later, his dead, heavily rotted body was found in the catacombs—not too far from where he entered the underground passageways. Philibert was perhaps the first of several people who would get lost in the city's catacombs, many of whom would never make it out.

The necropolis that lies beneath the City of Light was constructed in response to overcrowded aboveground cemeteries, such the Les Innocents, the city's oldest and biggest cemetery. So in the eighteenth century, specifically from 1786 to 1788, by the light of the moon, processions took place, shepherding the bones of the long-dead from cemeteries to the limestone mines that had long existed underneath the city. Eventually, the bones of six million people were deposited in various tunnels.

By the early nineteenth century, the underground ossuary, sixty-five feet below the streets, had become a tourist attraction. The Parisian upper classes were particularly curious. Even the Count of Artois, the future King Charles X of France, would descend into the depths below Paris to gawk at the bone depository. The public could visit the ossuary only with permission from a mine inspector. Then, in 1833, the church successfully lobbied to have the catacombs shut for good, reminding city officials about the taboo of exposing the living to the sight of dead bodies. But human beings' need for the macabre persisted. In 1850, the city allowed visits four times per year. Seventeen years later, they upped it to monthly. Then biweekly. Then weekly. Finally, in the early nineteenth century, the city opened up the catacombs to the public for daily visits. In World War II the French Resistance literally went underground in the catacombs to hide from the Nazi occupiers of the city.

There's a secret and illegal subculture in Paris, people called "cataphiles," who sneak into the off-limits swaths of Paris's vast underground network of

Paris's catacombs provide a haunting journey to the underworld for tourists and "cataphiles" alike.

caves. Sure, you can take a guided tour to the permissible parts of the catacombs. But some people with a yen for adventure go off the beaten catacomb path, finding the handful of secret entryways to the Parisian netherworld and wandering these oft-unmapped underground passageways.

For those who want to do it legally, there are guided tours, though only a mile of the vast subterranean network is open to the public. As visitors descend the stairway to the ossuary, they are met with an inscription above a doorway: *"Arrête, c'est ici l'empire de la mort!"* (Stop! This is the empire of death!). And it's not exaggerating. From

Signs throughout the Paris catacombs indicate from which cemetery the bones were reinterred, such as these moved on July 2, 1809, from the cemetery known as Les Innocents.

there, visitors stroll past piles and piles of human bones, many of them becoming ersatz walls flanking the miles of paths that vein the French capital. And they're not just stacked haphazardly. There are skulls and femur bones arranged to form columns in the middle of a room. There are skulls placed to form the shape of a cross. And if you just can't get enough, there's an Airbnb listing—a contest only available on Halloween night, for two "lucky" winners—that allows you to sleep in the catacombs among the skeletons. "Become the only living person to wake up in the Paris catacombs," it says.

Inside the Capuchin Convent's catacombs in Palermo, Italy, which hold more than 8,000 mummies.

IN SICILY THEY TAKE THE macabre to a whole new level. In Palermo's Catacombe dei Cappuccini, the Capuchin Catacombs, there are over 8,000 corpses and about 1,250 mummified bodies, many of which are on display for visitors to see.

Like in Paris, this southern Italian catacomb began when the aboveground Capuchin cemetery ran out of space to continue burials. So the monastery started carving out spaces in the crypts below the cemetery in the sixteenth century. They'd first dehydrate the dead monks on racks made from ceramic pipes and then either embalm the body or put it in an airtight glass box. For a while it was only the monks whose final resting place would be in this subterranean spot. By the nineteenth century, though, many upper-class Sicilians considered being entombed and mummified in the Palermo catacombs a status symbol and would write it into their wills, insisting their remains be dressed in certain outfits they liked.

As one could guess, the catacombs of Palermo have long been a tourist draw. After visitors head down a staircase inside the Capuchin monastery, there are halls and halls lined with mummified cadavers, each one dressed in monk's robes or their Sunday best. The halls are organized by sections for monks, women, men, virgins, and children. Some of the bodies look better than others, revealing that the old ways of embalming were still in need of advancement and progress. Walk down the passageways to see contorted or bloated cadavers, some looking like they're frozen in a half-deteriorated state. Others, in glass coffins, show some promise of preservation.

And then there's the case of Rosalia Lombardo. Just seven days shy of her second birthday, Rosalia succumbed to pneumonia. Her grieving father, Mario, took the body to a well-known embalmer, Alfredo Salafia, in the hope of perfectly preserving Rosalia in her current physical state. It was good timing for the mourning father. Salafia had been experimenting with a formula that he hoped would be superior to the methods used at the time. He came up with a cocktail of formaldehyde, alcohol, glycerin, zinc sulfate, and chloride to inject into the body of the child.

Rosalina Lombardo goes down in history not only as one of the last bodies to be placed inside the Capuchin Catacombs (in 1920) but also to be one of the best preserved, as her body has shown very little deterioration. That is until recently, when catacomb employees began to notice some discoloration of the skin. Rosalia was removed from her glass casket and placed in a new hermetically sealed glass box to help stop any further corrosion.

The Great Hall of the Bulls near Montignac, France.

LASCAUX
Caves

SOUTHERN FRANCE

LIKE MANY HISTORY-CHANGING UNDERGROUND DISCOVERIES, THE caves of Lascaux were first unearthed by an unsuspecting teenager. On September 12, 1940, eighteen-year-old Marcel Ravidat came upon a hidden cavern when searching for his beloved dog, Robot, whom he believed had fallen into a hole in the ground. He thought about descending into the hole himself but reconsidered and instead asked three friends to help. The four of them took the plunge and soon found themselves in a cave, discovering long-dormant spaces with walls and ceilings covered in paintings. There was one room, now called the Great Hall of the Bulls, plastered with images of horned male bovines; another,

The entrance to the Lascaux caves museum.

the Chamber of Felines, is, as one would expect from such a name, filled with cat drawings of various sizes.

The boys had found the now-famed caves of Lascaux. The cave chamber is just one of many ancient caves with Cro-Magnon–era paintings around France and Spain, but it's certainly the most famous.

When it comes to Lascaux, though, there's good news and bad news.

Let's start with the good news.

Located near the village of Montignac in the Dordogne region in southwestern France, Lascaux generated a dazzling amount of attention when it was first found. It suddenly added to the already-extant knowledge that our early ancestors were more than just grunting, fire-starting cavemen. It also showed us a completely different function of early art. Today we create art to display: it's a commodity. But twenty or thirty millennia ago, art had the opposite function, as we see from Lascaux. Probably very few people had access to these remarkable drawings buried deep in a cave. Many interpretations of the drawings' purpose have been put forward: that they were painted to commemorate a very successful hunt was an early theory. Beginning in the late twentieth century, a popular idea emerged from German scholar Michael Rappengluck: that the eyes of the bovines, birds, and birdmen represent stars and make up an ancient star map, one that would have directly mapped the stars as they were seen some fifteen to seventeen thousand years ago. According to this theory, three prominent stars are represented: Altair, Deneb, and Vega, known as the Summer Triangle because they are the most visible in the warm-weather months.

The cavern is covered with over two thousand parietal—or cave—paintings that are estimated to be about twenty thousand to thirty thousand years old, depending on who you ask. The complex is made up of a few long corridors, some of which expand to become larger chambers. And the scenes depicted aren't just primitive scribblings and doodles on the walls, but advanced paintings and drawings. The images, painted in red, black, and yellow, are largely made up of bulls and horses, but there is a smattering of felines, birds, and bears too. Oh yes, and a unicorn. There is only one image of a human, as human imagery at this time was extremely rare—or at least in the cave art humans have since discovered. And there's one rhinoceros image. The horned beast we associate with Africa actually roamed Europe until about ten thousand years ago. Perhaps the most famous images of Lascaux are found in the Great Hall of the Bulls. When you first enter the cavernous space, you see other animals: horses of various sizes, for example. But journey farther into the cave, and you can see how the room got its name. There are four large bulls depicted on the walls, one of which is seventeen feet long.

Picasso turned up at Lascaux not long after its discovery. "We have learned nothing in twelve thousand years," the painter supposedly said, when he was standing in front of the paintings.

Visitors gaze at a reproduction of the Lascaux cave art.

The animals that this prehistoric people painted on the cave walls were not necessarily the same animals they hunted. Reindeer was a common meal in this part of the world twenty thousand years ago, yet there are no images of them. Instead, it's been suggested that by emphasizing the strength and beauty of bulls and horses, the art serves a more symbolic, even spiritual, purpose. In this way, the cave is an ancient temple, serving not for shelter but for ceremonial purposes.

And so who created these marvelous images? Experts can only speculate, but they surmise they were painted by a few successive generations of people who passed down the painting technique to their offspring.

And now for the bad news.

What you're looking at when you make a pilgrimage to Lascaux to see the incredible cave art is not really the Lascaux cave art. Not exactly. It's a replica. The site was opened to the public in 1948, just eight years after it was discovered, and sits about six hundred feet away from the original. But by 1955, archaeologists and art historians noticed some deterioration taking place on the paintings. They closed it in 1963 for fear that the humidity, carbon dioxide, and body heat from the nearly twelve hundred visitors per day would destroy these priceless cave paintings. In fact, in the decade and a half after the cave opened to the public, the precious paintings deteriorated more than in at least ten thousand years. Sequels are rarely on par with the original, but the faux Lascaux, or Lascaux II as it's officially called, is still worthy of checking out. After all, the French government spent $67 million to ensure visitors would feel like they were looking at the real deal. The creators of Lascaux II, a cave which sits next to and mimics the original, went so far as to put speakers in the trees, emitting the sounds of the forest, so visitors can hear what the teenage boys might have heard when they traipsed through what once was a desolate wooded area.

> *What you're looking at when you make a pilgrimage to Lascaux to see the incredible cave art is not really the Lascaux cave art. Not exactly. It's a replica.*

Meanwhile, the original Lascaux continues to be plagued by the elements. A fungus was found around the year 2000, which probably developed from a new air conditioner that was installed. In 2006, researchers found black mold in the cave. For now, only one human goes in the cave once a week for just twenty minutes to check the climate conditions.

So the bad news is also the good news: thanks to the remarkably well-done replica, we will hopefully have the original around for time immemorial, even if only a select few humans will lay their eyes on it.

Churchill's WAR ROOMS

LONDON, ENGLAND

IN MAY 1940, BRITISH PRIME MINISTER WINSTON CHURCHILL STOOD IN a concrete bunker beneath central London and proclaimed, "This is the room from which I'll direct the war."

He was in the War Rooms, a newly constructed bunker that would serve as the nerve center for Britain's war effort in World War II.

Prime Minister Winston Churchill seated in the Map Room, which showed the position of every British convoy around the world.

Completed a week before Britain declared war on Germany, the War Rooms were originally created in response to air raids during World War I. Great Britain and London, specifically, were ill prepared for invading enemy air forces. In 1915 and 1916, Germans staged air attacks using zeppelins. They turned off the aircrafts' engines at eleven thousand feet above London, rendering them silent, and once over the city, they rained down a torrent of bombs on the city. The stealth attacks killed thousands and destroyed many historic buildings.

When the war ended, the British government began preparing a plan to defend against future conflicts. The government estimated that with the advancing technology in aircraft, future air raids could lead to the deaths of several hundred thousand citizens in just the first few days of an attack.

Among other preparations, they contemplated how to keep the government running in the event of sustained air raids. They considered evacuating the prime minister and other essential leaders out of the country but ultimately decided that the best plan was to have a bunker in which the government could continue to operate. They found an underground space used by the Civil Service to store furniture and began renovating it as the place where Britain would strategize for the war.

The War Room beneath Whitehall was originally vulnerable to a direct bombing, so Churchill personally oversaw it being recast in concrete.

And so, on August 27, 1939, the War Rooms were finished and ready to operate. And the timing could not have been better. On September 1, Germany invaded Poland. Two days later, Great Britain declared war on Hitler and Germany.

Out of the twenty-four rooms and eight spaces that are open to the public in the bunker, two are especially worth highlighting: the Map Room and the Cabinet Room. The Map Room was staffed around the clock by members of the Royal Navy, the Royal Air Force, and the British army, who collected important information about the enemy and possible future attacks for Prime Minister Winston Churchill and King George VI. The Cabinet Room was where the War Cabinet convened over one hundred times, the peak of which was during the legendary Blitz by German Wehrmacht planes.

The rooms were in use nonstop during World War II until August 16, 1945, when combat ended and the rooms went silent and dark for the first time since 1939. With the war over, the space was largely abandoned, though occasionally private groups of history buffs would arrange visits.

Then in 1984, the War Rooms were put under the jurisdiction of the Imperial War Museums, who apparently had much more zeal about the War Rooms than other branches of the British government. They promptly opened the underground bunker to the public. At the opening ceremony, Prime Minister Margaret Thatcher, a big admirer of Churchill, was in attendance. And in 2005, Queen Elizabeth II officially inaugurated the bunker as the rebranded Churchill Museum and Cabinet War Rooms (later renamed the Churchill War Rooms). Today the museum has many audiovisual and digital displays, including a forty-foot interactive table where visitors can access archived material and photographs simply by swiping their fingers on its surface. The museum also takes visitors through the life of Winston Churchill and has displays of his many military awards. Churchill fans might be especially gratified to see an exhibit featuring the prime minister's famous bowler hat.

Also on display are wall-size photos showing the destruction of London. Defused bombs hang from the ceiling. The Map Room, however, is largely unchanged, as if the war had just ended. Rows of rotary phones line the desks. Notes are pinned to the walls. Typewriters are loaded with paper, ready to print out a report.

In Churchill's private quarters, where he spent three nights during the war, is a bed with a comfortable-looking quilt and a desk with half-smoked cigars in ashtrays. He gave four wartime speeches from this room. There was no running water in the bunker, and so the prime minister was obliged to use a chamber pot inscribed with the crest of King George VI.

Churchill didn't like sleeping in the underground chamber, preferring instead to sleep at 10 Downing Street. When the war finally ended, he certainly breathed a sigh of relief that the death and destruction of war was, for the time being, a thing of the past. And it was undoubtedly a pleasure knowing he wouldn't have to spend any more time underground.

"Churchill didn't like sleeping in the underground chamber, preferring instead to sleep at 10 Downing Street. When the war finally ended, he breathed a sigh of relief that he wouldn't have to spend any more time underground."

HOME DEFENCE MAP
LEGEND

Churchill's underground bedroom
contained a BBC microphone, where
he made four wartime broadcasts

The entrance to Mortimer's Hole lies at the foot of Nottingham Castle. Tourists can also visit the Brewhouse Yard Museum of Nottingham right next door.

MORTIMER'S HOLE

NOTTINGHAM, ENGLAND

SOMETIMES UNDERGROUND TUNNELS LIE DORMANT FOR CENTURIES. Other times they remain in use over centuries or even millennia. And sometimes they change history—or they at least tell a historical tale of *Game of Thrones* proportions. Such is the case with the tunnel known as Mortimer's Hole, located in the city of Nottingham, about 120 miles north of London.

In order to fully understand the significance of this once-secret underground passageway, we must first head to France to meet Roger de Mortimer. Mortimer faithfully worked for the English king Edward I and had fought under the king's name in Ireland and Wales, pushing back the powerful forces of Edward Bruce (brother of the king of the Scots, Robert Bruce). But Mortimer's fortunes took a turn with the death of King Edward I. He did not get along with the heir, Edward II, and was increasingly unsatisfied with the new monarch's policies. After Mortimer led an unsuccessful revolt against Edward II, the king had him thrown into the Tower of London in 1322. But Mortimer escaped by drugging the guards and fleeing to France.

This is where things get really scandalous. It was in France that he met Edward II's Gallic wife, Queen Isabella. She didn't like Edward II, her husband, either. Possibly because she was at odds with him politically. And possibly because he was purportedly homosexual. She cooked up a reason for Edward II to let her go on a diplomatic trip to France to speak to her brother, King Charles IV. Not long after her arrival, Isabella and Roger began having a torrid love affair. United against Edward II, they set sail for England with a large mercenary army from France to depose the king. Edward II fled

One of the nearly 550 medieval tunnels carved from sandstone underneath Nottingham Castle.

to Wales but was nabbed under the orders of Isabella. She sentenced him to life in prison, but not long after his incarceration, he died. Historians disagree on his ultimate fate. Some say he succumbed to a red-hot poker in the anus—due to his suspected homosexuality—while others believe he simply died of bad health due to the terrible conditions of his captivity.

Though Isabella's son, Edward III, became king, it was Isabella and Mortimer who quietly ruled England. Edward III was frustrated by the power that Mortimer wielded, however, and on October 19, 1330, the seventeen-year-old Edward III and a legion of supporters turned up at the castle in Nottingham, where Isabella and Mortimer were ensconced. The townspeople of Nottingham told the young king and his men about a secret passageway—the tunnel now known as Mortimer's Hole—and they marched through it and right into the fortress.

Hearing a ruckus in the castle and suspecting it was her son Edward III, Isabella burst out of her bedroom and screamed, "Fair son, fair son, have pity on gentle Mortimer!" The son did not obey his mother's orders. Mortimer was apprehended and whisked away via that same underground tunnel. A few days later, he was back in the Tower of London, imprisoned and awaiting his fate. That fate came on November 29, 1330, when Mortimer, branded a traitor, was hanged. His body was left to hang there for two days so everyone could see the spectacle. Eventually, Mortimer's corpse was interred less than two miles away at Church of the Greyfriars.

With his reign finally secured, Edward III put his mother, Isabella, under house arrest at Castle Rising in Norfolk, 110 miles north of London. She remained there for thirty-one years, and in the process went mad. Upon her death, she too was interred at the Church of the Greyfriars in London.

There are almost 550 tunnels that run underneath Nottingham Castle, and many have historically been used as storage rooms, dungeons, and (more recently) bomb shelters. But the most famous, of course, is the passageway associated with our dear Mortimer. Today visitors can take part in the intrigue with a walk through Mortimer's Hole. If you take a guided tour, you can hear all the sordid details of this tragic-yet-entertaining tale. The tunnel, about 320 feet long, begins in the Brewhouse Yard.

Or does it? In 2010, archaeologists made a startling discovering while surveying the sandstone caves. The passageway historians thought to be Mortimer's Hole was one regularly used for carting things up from the river to the castle—so not so secret after all. But archaeologists now think they've been focusing on the wrong tunnel for centuries. It is believed that the true Mortimer's Hole is a tunnel called the North Western Passage—a truly secret and rarely used tunnel.

> *The passageway historians thought to be Mortimer's Hole was one regularly used for carting things up from the river to the castle—so not so secret after all.*

And how do the authorities at Nottingham feel about this discovery? In a BBC report, Dave Green, who runs the heritage sites for Nottingham City Council, said, "History is always controversial and full of differing opinions and ideas." He then added, "We will look forward to presenting this new information alongside the stories we have always told on our cave tours and leave for the public to choose for themselves which is the real Mortimer's Hole."

Whichever tunnel you find yourself in when visiting Nottingham, think of the historical intrigue that took place here nearly seven hundred years ago, as you imagine Edward III's cronies whisking Mortimer away to his eventual demise.

An eerie view of Mary King's Close.

EDINBURGH, SCOTLAND

BURIED BELOW THE ROYAL MILE, ONE OF THE MAIN THOROUGHFARES of Edinburgh's Old Town, is a hidden street that is dark in more ways than one.

The Royal Mile is intersected by tiny alleyways that lead off to the north and the south. These smaller lanes, known as "closes," are flanked by six- or seven-story buildings, creating a canyon of very dark and dingy passageways. In the seventeenth century, these closes were some of the most bustling parts of Edinburgh, as traders sold merchandise and people shopped on the streets. Legions of humanity were housed in the cramped buildings, with families of up to twelve people living in a single room. The steep-angled lane of Mary King's Close pointed toward old Nor Loch, where there was a sewage-filled marsh that was a common spot for drowning suspected witches.

As if this were not unpleasant enough, things got worse in the middle of the seventeenth century. In 1645, disease-carrying rats arrived on cargo ships, spreading the bubonic plague to humans and killing off about a third of the city's population. The Scottish parliament even moved itself out of town to avoid it. In order to try to stop the plague from spreading, the city made new laws: any house that had previously sheltered infected people would be under quarantine, and those exposed would be sent to quarantine camps outside the city walls. Edinburgh's plague doctor George Rae would be called to help victims in his birdlike mask and long overcoat, although there was little he could do beyond bursting the buboes and cauterizing the wounds left behind. Victims' homes were then cleansed by specially appointed "foul clengers," who wore gray tunics marked with a white saltire—the cross of Saint Andrew, Scotland's patron saint. Goods, bedding, clothing, and even roofs were usually burned. Smoke from the flowering shrub broom was used to cleanse the air. Though no one in Edinburgh was safe from the plague, the disease was particularly effective in spreading around the city's damp, narrow closes.

In the mid-eighteenth century, the buildings of Mary King's Close were deemed unfit for human habitation, and the city built the Royal Exchange on top of the upper part of the close. However, residents remained in the lower homes until the late nineteenth century, with the saw-making workshop of Andrew Chesney remaining in use until the year 1901.

In response to numerous reports of apparitions and other alleged paranormal behavior, a Japanese psychic, Aiko Gibo, visited Mary King's Close in 1992. Sensing the pain and misery that had taken place here, Gibo could barely step inside one of the rooms. The psychic felt an almost physical presence of people around her, huddled in blankets, sad and ill.

The psychic felt an almost physical presence of people around her, huddled in blankets, sad and ill.

As she was leaving the room, she felt a tug on her pant leg and turned. Standing in front of the window was a small girl with long, dirty hair, dressed in rags. She was crying and told Gibo she was searching for her mother, who had left her there during the plague, and that she yearned to have a doll to keep her company. Since then, people bring all manner of toys for the girl, now known as Annie. Today visitors can see them amassed in a little nook in Annie's room. Although extensive research has been carried out, no child by that name appears to have lived there.

Mary King's Close reopened to the public in April 2003 as a tourist attraction. If you are brave enough to descend into the space, you'll enter through the exhibition center on the Royal Mile and let your character guide, portraying a onetime resident of the close, lead you through the warren of streets, eerie low-ceilinged homes, and old shop rooms, regaling you with tales of actual residents of the close. Learn about life in seventeenth-century Edinburgh, such as early sanitation practices, which involved tossing the contents of chamber pots out the window twice per day at seven in the morning and ten in the evening. When a resident would send an object raining down into the street, she or he would yell "gardyloo," a corruption of the French *gardez l'eau*, which means "look out for the water!"

Centuries ago, the area was subjected to marsh gasses. Experts say that visitors may see strange, vague light formations or even hallucinate because of the lingering effects of the gas. Or maybe visitors just don't want to admit to themselves they've seen an actual ghost.

Mary King's Close was closed to the
public from 1901 to 2003.

The Rynek Główny, Kraków's medieval square.

RYNEK Underground

KRAKOW, POLAND

VISITORS RARELY CONSIDER WHAT IS UNDERNEATH KRAKÓW. That's because Poland's gem of a historical metropolis is filled with striking attractions, including Wawel Castle; Saint Mary's Basilica with its incredible carved wood altarpiece; the old Jewish-neighborhood-turned-hipster-haunt, Kazimierz; and, of course, the Rynek Główny, the city's thirteenth-century main square surrounded by elegant buildings of various architectural styles and its former trading center, Cloth Hall, with its Gothic arches and clock tower.

But lying just below the Cloth Hall is something to rival Kraków's street-level charm. The only indication there's something to see belowground is a small plexi-glass pyramid, lit up in blue, on the main square.

But you could learn as much or more about Kraków from its main underground attraction than by strolling its cobbled streets. The Rynek Underground, first opened to the general public in 2010, is a vast high-tech museum that takes visitors through Kraków's fascinating history. It's the size of three football fields. As one of Central Europe's major metropolises in medieval times, Kraków was a mecca for trade. Everything came through here. Descend thirteen feet below the cobblestones of Rynek Główny to explore the ancient merchant stalls, where there's a nearly 43,000-square-foot space filled with films, interactive displays, and historical artifacts that bring the city's history to life.

Amazingly, it wasn't until 2005 that city authorities thought much about what might lie below the town's main square, one of the most attractive and largest medieval public spaces in Europe. Originally, they planned a six-month archaeological dig. Archaeologists began to find all manner of priceless historical mementos, including articles of clothing, coins, military weapons, and jewelry. Then came the major finds: remnants of an aqueduct, a cemetery, and streets from seven hundred years ago that hadn't been traversed in centuries. It was incredible objects like these that turned the dig into a five-year project, as historians and archaeologists were eager to discover what else could be down there.

When visitors step out of the elevator, emerging underground, the first thing they encounter is a wall of semitranslucent fog with a medieval street scene projected onto it, an attempt to take visitors seven hundred years back in time to medieval

The Rynek Underground, a well-preserved medieval public space in Kraków.

Kraków. A hologram of a trader appears, and he's angry with the visitors, screaming in Polish that they don't have an official license to sell bread. The figure's vaguely opaque form makes it seem like the action is actually taking place on the street in front of you. Walk over a glass footbridge and peer down at the medieval streets below, complete with ankle-breaking cobblestones.

The museum's designers took advantage of the well-preserved space. Each of the underground's original stone niches, for example, is outfitted to look like a medieval shop. Since the city's main square has long been a hotbed of trading activity, the curators attempted to emulate what went on here in the past. There are displays of historical jewelry, coins, horseshoes, knives, shoes, and padlocks and keys, among many other artifacts.

The museum also boasts holograms, lasers, numerous touch screens, smoke machines, and the artifacts that were recovered in the dig, including objects from other parts of the world—physical evidence of Kraków being an important trade center in the Middle Ages. Children can have fun weighing themselves with an old hundred-weight. There's a fascinating audiovisual display of what the city was like when it was raided and plundered by the Tatars in

One of the most intriguing parts of the museum is an explanation about the proper burial of a vampire. For those who want or need to know, here it is: bury the body in a fetal position with the head cut off and positioned between its feet.

1241. A detailed scale model of the sixteenth-century city gives visitors an idea of how Kraków has—or has not—changed in several hundred years. Above it is a large skylight with a fantastic view of Saint Mary's Basilica. It's all designed to give visitors a better sense of what daily life was like in medieval Kraków.

There is also an old cemetery in which the bodies of several suspected vampires were found. One of the most intriguing parts of the museum is an explanation about the proper burial of a vampire. For those who want or need to know, here it is: bury the body in a fetal position with the head cut off and positioned between its feet; or, in other cases, place sickles around the neck and large rocks in the mouth and on the body to weigh down the corpse, which, according to the medieval mind-set, could help prevent reanimation.

After such an experience, you may need to retire to an aboveground pub for a shot of bison-grass vodka, a particularly Polish embodiment of one of the country's favorite spirits. And perhaps eat some garlic.

Just in case.

Ksiaz Castle conceals a network of underground Nazi tunnels in Wałbrzych, Poland.

KSIAZ CASTLE

WALBRZYCH, POLAND

IN AUGUST 2015, AMATEUR TREASURE SEEKERS ANDREAS RICHTER AND Piotr Koper were scanning a swath of land in Lower Silesia with a ground-penetrating radar when they noticed something: an object under the ground. It was large and rectangular. They speculated that it could be an oversize vehicle of some kind. Could it be a train? Could it be the discovery of a lifetime?

Soon enough newspapers were reporting on the possible find. Tourists sprinted to the site, located near Ksiaz Castle outside of the town of Wałbrzych, Poland.

So what was all the fuss about? The object Richter and Koper detected sitting beneath the surface of the ground might have been the near-mythical Nazi gold train that history buffs and treasure hunters have been talking about since the end of World War II.

Near the end of the war, this part of Central Europe belonged to Germany (it would, after the war, be transferred to Poland, and all Germans would migrate westward across the new border). In early 1945, many of the German-speaking residents of Lower Silesia, hearing that Soviet soldiers were advancing toward Germany from the east, hid their valuables in the ground and left. It's been whispered by treasure hunters that mountains of gold and heaps of priceless jewelry must be hidden underneath the area.

But in addition to the rumored valuables of long-departed residents, there is, located beneath Ksiaz Castle, a subterranean enigma wrapped in a Nazi mystery. In 1943, the German military turned up at the castle and undertook a massive underground construction project. The Germans initially brought in a few hundred Italian workers and coal miners to start the dig. But by 1944, there were three thousand people working underneath the castle—all of them prisoners of war and detainees from nearby concentration camps. (Eventually up to five thousand forced laborers would die from the harsh conditions.) The end result was a massive tunnel that spans eighteen feet wide and sixteen feet high. Work went slowly because the rock was gneiss, a strong and sturdy stone. The tunnel is part of a large network of subterranean passageways called Project Riese, the German word for "giant."

The complex covers about forty-six acres of land and includes seven different subterranean spaces, all of which were eventually going to be connected by tunnels. The other nearby locations include Rzeczka, Włodarz, Osówka, Sokolec, Jugowice, and Soboń. The tunnels are in various stages of completion, as some look like crude passageways and others are clad in smooth concrete, almost ready to be used.

The ambitious project was never fully completed, and by the end of the war, the Germans closed up the tunnels and fled.

Ksiaz Castle was at the center of this not-completely-realized project. Built at the end of the thirteenth century, the four-hundred-room fortress is one of the biggest of its kind in Europe. It was long under the ownership of the powerful noble Hochberg family, and famous people who have visited include Czar Nicholas I and US president John Quincy Adams. The last Hochbergs to reside there were Hans Heinrich XV, Prince of Pless, and his British-born spouse with a mouthful of a name, Mary-Theresa Olivia Cornwallis-West. The Hochbergs fell into debt, and in 1941, the Nazis took over the castle, some say as revenge for the two Hochberg sons who were fighting for the British army and the Polish army, respectively.

With the castle now occupied by the Nazis, the new residents went ahead and renovated the place. Gone were the lavish baroque adornments and furniture. In their place was a stark, cold Nazi-friendly interior—offering a hint that the castle was meant to be a future home of the Führer himself.

But the amazing thing about Project Reise is that still to this day no one really knows what the Nazis' intentions were with it. Some people have speculated it was supposed to store airplanes. Others think it was meant to be an underground city, a bombproof shelter, for Hitler and other Nazi elite. And that would be a lot of Nazis. It was estimated that it could hold twenty-seven thousand people. But ever since the end of the war, treasure seekers have been convinced this is where the Nazis stored the gold that they'd looted far and wide. One other theory, inspired by the tunnels having been used by German scientist Hubertus Strughold, who would later be called the "Father of Space Medicine," for research during the war, is that the Nazis had been storing a spaceship they had built in part of the tunnel system.

Today visitors to the castle can tour one of the tunnels that run underneath the fortress. There are two chambers located directly below the castle courtyard, one 50 feet and another 160 feet down, but only the former is open to visitors. The deeper chamber is occupied by the Geophysical Observatory of the Polish Academy of Sciences. The three hundred or so feet of the underground area that is open to the public is mostly just spartan today, its concrete walls are about the only thing to look at. It's an empty space filled with the air of mystery. But you can use your imagination to try to guess what the Nazis' intentions were.

As for Richter and Koper's gold-filled train, that turned out to be a false alarm.

And so the mystery remains, and the treasure hunters of Lower Silesia continue to comb the countryside of southwest Poland, hoping they'll stumble upon that Nazi gold.

A section of an unfinished tunnel reputedly built by the Nazis during World War II to transport gold.

TUNEL SPASA · KUĆA KOLARA

ULICA
TUNELI

1

The Tunel Spasa house where the entrance of the Sarajevo War Tunnels was accessed during the siege of Sarajevo, Bosnia.

SARAJEVO War Tunnel

SARAJEVO, BOSNIA, AND HERZEGOVINA

THE TWO-STORY HOUSE IN BUTMIR, A NEIGHBORHOOD NEAR Sarajevo's airport, looks ordinary enough. Its unpainted brick-and-mortar facade is riddled with pockmarks, perhaps scars of the Bosnian conflict that raged from 1992 to 1995. Blemishes and all, it looks like any other house on the unassuming block.

This is, though, no ordinary house. The structure contained a secret. It was the entranceway to a clandestine tunnel that stretched 2,625 feet to the neighborhood of Dobrinja. The reason? The siege of Sarajevo.

The fighting began in early April 1992, when tanks rolled into Sarajevo. It lasted until February 1996, the longest siege on a capital city in modern times. Yugoslavia was splitting apart, and it was doing so along ethnic and religious lines. Bosnia, with its population of Muslims (referred to as Bosniaks), Serbs, and Croats, was like a Yugoslavia in microcosm. And just like in the wider region during the Balkan wars of that decade, in Bosnia it was the Serbs who were largely the aggressors. In the case of Bosnia, in general, and Sarajevo, in particular, it was the forces of Republika Srpska, made up of Bosnian Serbs, that had had cut off Bosniaks living in Sarajevo from the rest of the country. They set up roadblocks with tanks. They positioned snipers in the mountains. They were constantly on patrol to make sure no Bosniaks could leave their sequestered part of the city.

Bosnian-Serb checkpoints prevented the residents from importing necessary goods into their part of Sarajevo. They feared they would starve, even though everyone was rationing their food. But there was one Bosniak-dominated neighborhood that had a possible loophole: Butmir. It had access to free Bosnian territory. There was just one problem: the city's airport, specifically the runway, was between Butmir and the rest of the contained, Bosniak part of Sarajevo. And sprinting across the runway wasn't an option. Even though the airport was officially controlled by the United Nations —the UN had negotiated this with the Serbs so they could bring in humanitarian aid— Bosnian-Serb snipers had their guns ready at all times.

So, there was only one option: dig a tunnel. The effort began in January 1993. Residents of each neighborhood were given a shift, and work crews began laboring around the clock, digging a subterranean passageway that started on either side of

Constructed in 1993, the Sarajevo War Tunnel was used to transport supplies, food, and weapons during the war.

the runway. It spans from one home's garage in Dobrinja to the aforementioned unassuming house in Butmir. Workers from the two neighborhoods dug toward each other, hoping they'd meet somewhere below the runway. The code name of the project became Objekt BD—the first letter of each neighborhood, Butmir and Dobrinja, that the tunnel connected. They worked in eight-hour shifts, and laborers were paid with one pack of cigarettes per day, which was a hot commodity during the siege, as they could be traded for other necessary items.

One of the main issues during the dig was leaked water filling up the tunnel. Sometimes it would rise up to the waist of the workers. There was also no ventilation in the tunnel, so workers had to wear masks. And the only sources of light for the laborers were "war candles," containers of cooking oil with a string for a wick. Every minute of every day workers shoveled and picked at the tunnel, steadily making their way toward each other under the runway.

On July 30, 1993, six months after starting to dig, the two crews met in the middle. The tunnel, which stretches a half mile, is five feet high and three feet wide. They added wooden support beams and eventually a narrow-gauge railway so that food, equipment, and weapons, among other supplies, could go from point A to point B much quicker. It has been suggested that more than a million trips were made in the tunnel after its completion—and that up to twenty million tons of food were transported via the underground passageway. They also ran a high-voltage electricity cable and a pipeline for oil to the cut-off sections of the city. The tunnel allowed the Bosniaks to survive. It was literally a lifeline.

It has been suggested that more than a million trips were made in the tunnel after its completion— and that up to twenty million tons of food were transported via the underground passageway.

When the war was over, the owner of the house in Butmir, Bajro Kolar, turned it (and a part of the tunnel) into a museum, the Sarajevo War Tunnel Museum, which gives a fascinating account of how the passageway was made. A looping eighteen-minute film shows footage of the siege of Sarajevo. Some of the digging equipment that was used is on display, as well as military uniforms and wartime photos.

After the war, with the tunnel no longer a necessity, it eventually caved in, thanks to the constant pressure of groundwater.

But visitors to the Sarajevo War Tunnel Museum have the good fortunate to access about eighty feet of the preserved passageway that runs underneath Mr. Kolar's house.

The tunnel is now nicknamed the Tunnel of Hope.

BUNK'ART

70 vjet pas çlirimit

The exterior of Bunk'Art, one of several former
bunkers transformed into progressive art galleries.

BUNK'ART

TIRANA, ALBANIA

THE SOUTHEAST EUROPEAN NATION OF ALBANIA IS SO OBSCURE that in the 1997 American film *Wag the Dog*, a PR firm was hired by the American president to invent a fake war with Albania to get the media's attention away from a presidential scandal. Why Albania? Because no one knows anything about it, and most Americans can't locate it on a map. It was the perfect foil at the time.

Albania takes the prize for being one of the most isolated countries of the twentieth century. And it's all due to the nation's bunker-loving leader, Enver Hoxha (pronounced *Ho-ja*).

Hoxha, who ruled the country from 1944 to 1985, was so paranoid of foreign invasions, he went on a bunker-building spree in the 1970s, placing over 700,000 igloo-shaped bunkers everywhere: on towering hillsides, in city parks, in the middle of vast fields, outside of obscure villages. In addition to dotting the landscape with underground shelters, Hoxha had electric fences topped with barbed wire installed near the border and declared 64 percent of the country's gorgeous Adriatic coastline a military zone. No one ever invaded Albania, and today the legion of bus-size fortifications remain, relics and symbols of a time most Albanians would prefer to forget.

But there was one bunker he kept secret. It's a massive 106-room, hidden concrete fortress from the 1970s that is built into a hill on the outskirts of the capital, Tirana. The five-floor complex, which goes as deep as 330 feet into the ground, was to be home to Hoxha and his cronies in case the United States (or any other country) really did go to war with it.

Inside the entrance to the Bunk'Art, which was once used as an antinuclear war bunker and has a new life as an art space.

Fortunately for everyone on the planet, but especially the people of Albania, Hoxha never had to use the bunker as an actual bomb shelter. In fact, he died in 1985, a couple of years before it was even completed. Hoxha left Albania as the third-poorest country on the planet, with the per capita income of the average Albanian at just fifteen dollars per month. The country was in bad shape. Construction on the bunker was completed, and then it was sealed up and forgotten about, much like the country itself. That is, until recently. The city of Tirana is doing something intriguing with the once-secret fortification.

It all began in 2014, when the bunker, now known as Bunk'Art, opened up its thick metal doors to the public. Inside, forty of the one hundred rooms are art exhibits—particularly avant-garde art that was banned under Hoxha—and displays explaining the country's history under the leader. Chronological time lines take visitors through the history of twentieth-century Albania. The space not only reminds Albanians of the tyrannical rule and iron grip of Hoxha and the Communist period but also encourages many Albanians to have a dialogue about the past, a moment of reconciliation.

In addition to isolating the country, the paranoid Communist dictator and one-time Stalin mentee is said to be responsible for the deaths of at least five thousand people, the imprisonment of over twenty-four thousand Albanians, and the displace-ment of seventy thousand during his forty years atop his red pedestal. There were up to two thousand informants living among ordinary people. The secret police held files on the personal lives of one million Albanians—and this was in a country of just three million people.

In the first two months it was open to the public, over sixty-five thousand people poured in, most of whom were Albanians. Sure, they wanted to see a historical curiosity that hitherto had been unknown to the general public. But it suddenly forced an unexpected conversation within Albanian society about the Communist past.

Stroll through the cold, quiet hallways, buried deep in the mountain, and pop in to various rooms. You might see rooms filled with creepy-looking gas masks. Rooms displaying old military uniforms. Rooms showing a chronology of Albanian history since the Fascist Italian invasion of 1939. Rooms with photos of a handful of those three-quarters of a million bunkers sprinkled around the country. Rooms showing conceptual art on old TVs, the meaning of which may be hard to glean.

And then there's the massive assembly hall, which would have been used for the entertainment of top Communist leaders and for Hoxha to address the bunker dwellers all at once. In 2016, the group that runs Bunk'Art staged a jazz festival here—jazz being one of the art forms that Mr. Hoxha had banned during his rule, of course.

One of the most memorable rooms in the complex is Hoxha's office: the center-piece of the wood-paneled room is a sturdy wooden desk holding what looks like a radio from the 1950s and an ashtray. Pick up the old rotary phone, and you'll get to hear a looped recording of Hoxha speaking in the earpiece. The multicolor armchair would not have been out of place in someone's 1970s rec room.

Bunk'Art, despite being on the periphery of town, is definitely worth a visit. And if you don't feel like making the journey to the outskirts, the city has also opened Bunk'Art 2, which is smack in the center of Tirana, located just under the central Skanderbeg Square. Thanks to the popularity of Bunk'Art, organizers decided to transform what was another secret underground shelter. With five-foot-thick concrete walls, the sequel, like its predecessor, exhibits art and historical displays. The bunker, built from 1981 to 1986, belonged to the Ministry of Internal Affairs. A new entrance had to be built to open to the public—the original one could only be accessed by the ministry. The "Pillar," as the museum is called, named for the Cold War–era government code name for the underground space, has a main permanent exhibit featuring photographs and documentary footage of the tens of thousands of ordinary Albanians who were persecuted under Enver Hoxha. Visitors can wander into twenty rooms that show images involving the country's secret police and hold the names of the thousands of people who were killed or imprisoned during the period.

It's just one more step—one more step underground—for Albania to come to grips with an awful period in the nation's history.

EATING & DRINKING Underground

CHEZ MOI, New York, USA

CAVE BAR, Dubrovnik, Croatia

RISTORANTE GROTTA PALAZZESE
(The Seaside Restaurant), Italy

MADAME CLAUDE,
Berlin, Germany

THE IDEA OF EATING AND drinking underground makes us think of speakeasies or private supper clubs. But as you'll see, anytime there's a cave or cellar or old train tunnel that is habitable enough to spend a few hours in, humans have made the best of it, wining and dining down under.

Beneath Brooklyn's bustling Atlantic Avenue lies a secret bibulous hideaway. To find it, though, you have to know where the hidden door is. Enter through the French restaurant Chez Moi and look for the bookcase, which, of course, is not a bookcase at all, but a doorway. Descend the dozen or so steps, and you're in Le Boudoir, an eighteenth-century-themed cocktail bar that is meant to evoke the private quarters of Marie Antoinette at Versailles. The rococo stylings include gilded mirrors, plush red banquettes, and antiques sourced from various châteaux in France. The door handle to the restroom is straight from Versailles itself, complete with an *MA*—Marie Antoinette's initials—engraved on the handle. Once inside, the flourishes become even more Gallic as the bathroom is meant to be a complete replica of Madame Antoinette's toilette.

The once undecorated space was discovered by accident. The owners of Chez Moi had heard rumors of the existence of some underground tunnels that ran beneath the restaurant. And so one night, after a couple of drinks, one of the

Le Boudoir, the subterranean bar in New York City's Chez Moi restaurant.

proprietors took a sledgehammer to one of the walls and voilà! They discovered two subterranean lairs, which they later learned were part of the oldest subway tunnels in the world. This tunnel, dating from 1844, wasn't connected to the famed New York City subway—that wouldn't be built until 1903. The tunnel that Le Boudoir now partly occupies was built to take the regular steam engine off the streets of Brooklyn, then an autonomous city, because of injuries that had been happening with pedestrians.

Fortunately, you can eat more than just cake here. Sip a French 75 and graze on some foie gras and crispy frogs legs while taking in this Gallic underground wonder eleven feet below street level.

TO FEEL LIKE A WINE-SIPPING troglodyte, head over to the historic town of Dubrovnik on the southern Dalmatian coast in Croatia. Until recently there was a wine bar in a Bronze Age cave directly under the airport runway. The subterranean space was first discovered just before World War I and was used as a bomb shelter during the Croatian War for Independence in the early 1990s. In 2014, the airport opened the grotto as a bar but closed it a couple of years later. The space is now only used for private events, but fortunately for people who love imbibing in a cave—and who doesn't?—the coastal city has a second bar in a grotto: about three miles from the walled Old Town is the aptly named Cave Bar More, located in the More Hotel (pronounced *Mor-ay*).

The cave wasn't actually discovered until the seaside hotel was under construction. That's when the proprietors

The multilevel Cave Bar More in Croatia.

decided to incorporate the grotto into the hotel and the Cave Bar was born. After you take the elevator down to the ground floor, walk straight to hit the beach or turn right for your cave experience. Sip on a glass of crisp white wine from the nearby Pelješac peninsula or nurse some rakia, a potent grape brandy that is popular in the region, and stare up at the stalactites—often illuminated in purple or red lights—pointing down from the tall ceiling. They intentionally play no music at the bar, so that the sound track is the ebb and flow of the sea, just outside the cave door.

Then raise your glass in a toast and say *Živjeli*—pronounced *Zheev-yelee*—"to life," just as the locals would.

MOVING ACROSS THE Adriatic Sea, near the southern Italian town of Bari, is the Ristorante Grotta Palazzese, a restaurant housed in a hotel of the same name. Here you can dine in a limestone cave, its large, open mouth looking out onto the Adriatic Sea, with the waves crashing onto the cliffs about seventy-five feet below. During the day, much of the light is natural, bouncing off the glimmering sea and lighting up the interior of the cave. At night, though, small lamps go on, creating a different but equally alluring ambience for the restaurant.

The limestone space has been no stranger to debauchery in the centuries past. In the eighteenth century, the ruling aristocratic family of the area would regularly throw parties inside the cave. It became so renowned with European blue bloods, in fact, that many young aristocratic men doing the grand tour around Europe would specifically stop by here, just south of Bari, to have their own *festa* inside the grotto.

People gravitate here more for the atmosphere than the food, which is on the pricey side—probably the most money you'll ever pay for dinner in Italy. That said, the tagliatelle with sea urchin, plus the incredible view and atmosphere in the cave, is well worth the pasta dish's fifty-dollar price tag. And because of the unique aspects of the grotto restaurant, reservations are a must.

BUT IF EATING AND DRINKING in a World War II–era brothel is more your speed, Central Europe is the place for you. In the gritty part of Berlin's Kreuzberg neighborhood, step down into Madame Claude, a subterranean dive bar that was once an underground brothel and a favorite hangout for American soldiers who were stationed in Berlin just after the war's end, when the German capital was quartered off between the French, British, Russians, and Americans. The bar, started by three French friends, is made up of several rooms. And if you notice something very "off" about the space, it

Italy's Ristorante Grotta Palazzese.

might not be because you're too intoxicated: Madame Claude is an "upside down bar," meaning all the furniture—chairs, tables, benches—is on the ceiling. And just to take the illusion one step further, some tables have everyday objects affixed to them, such as containers of aspirin, magazines, and votive candles. The reason for this Daliesque touch? When one of the current owners was a kid, his dream was to be able to walk on the ceiling. So at least this way, he has the illusion he's walking or—as Lionel Richie crooned—dancing on the ceiling. Have no fear, though: there are still chairs and tables on the actual floor, so you can kick back and enjoy your bottle of German beer.

Madame Claude's, a belowground—and topsy-turvy—bar in Berlin, Germany.

From the
DEAD SEA
to the
CASPIAN SEA

The Pool of Arches, also known as Saint Helen's Pool and Bir al-Aneziya, in Ramla, Israel.

THE POOL of ARCHES

RAMLA, ISRAEL

AMPLE NUMBERS OF TOURISTS IN ISRAEL THESE DAYS MAY CRUISE right past Ramla, a small town about halfway between Gaza and Jerusalem, near Ben Gurion Airport. It's been described as an "eyesore," which it very well may be. But lying beneath the town is one of the world's great cisterns and a structure significant to the development of Islamic architecture and design.

Ramla was founded in the year 716 by the Umayyad Caliph Sulayman, the name of the town being derived from the Arabic word for sand, *raml*. The Roman-era capital of the area was a coastal town called Caesarea, but after the Arab occupation in the eighth century, they moved the capital, first to nearby Lod. That didn't last long before they moved the capital again, this time to the relatively undeveloped Ramla, as it was located on the crossroads of a popular trading route between Cairo and Damascus. Because it was about ten miles from the coast, it would not be susceptible to attacks from the Byzantine navy. Little Ramla became known as the capital of Jund Filistin, one of the Palestine military districts of the Umayyad and Abbasid caliphates. Ramla quickly became a grand spectacle, as its streets were flanked with plus-size mosques, awe-inspiring administration buildings, massive mansions, and fountain-centered gardens. It boasted a diverse population of Muslims, Christians, Samaritans, and Jews.

The reason for ancient Ramla's disappearance to almost all but the history books is twofold: an earthquake in 1068 destroyed much of the existing structures; the city

Constructed by caliph Haroun al-Rashid in AD 789, the Pool of Arches features inner walls engraved with Old Arabic script.

was rebuilt in 1072, but it never regained its significant status. In the twelfth century, crusaders from Europe occupied Ramla and moved the center of town to the southeast, which is still today the hub of modern Ramla, leaving the old center, including the amazing cistern, largely neglected.

So if you want a clue to Ramla's former glory, head underground, where you'll find the only structure still remaining from the city's golden age: the Pool of Arches. The cistern actually has more than one name, depending on who you ask. Others call it the Pool of Saint Helena, a reference to the mother of Emperor Constantine, the Roman ruler who legalized Christianity in the empire. Helena, who is famous for the fourth-century sanctified shopping spree in the Holy Land that brought armfuls of Christian relics back to Rome, is said to have ordered the digging of the hole that eventually became the pool. Finally, some call it the Pool of Goats, because it was here that shepherds often took their goats to drink water.

Built in AD 789, this ancient cistern, seventy-eight feet by sixty-seven feet, supplied water to the twenty-five thousand denizens of Ramla. The soaring ceiling is made up of six vaults that are supported by five rows of three columns each. Before the cistern was built, the city had a reputation for bad drinking water. When the famed tenth-century Arab geographer Muhammad Ibn Ahmad Shams al-Muqaddasi visited Ramla, he wrote of precistern Ramla and its water issues in his book *The Best Divisions for Knowledge of the Regions:* "The wells are deep and salty . . . the poor go thirsty and strangers are helpless and at a loss what to do." So an aqueduct was built that connected Gezer Springs, about eight miles east of Ramla, with what became the Pool of Arches in the center of the city.

After Ramla fell into disrepair following the 1068 earthquake, the cistern was relegated to a place where residents would fetch rainwater that had collected in the basin. It eventually was filled with dirt. In the 1960s, though, Ramla town authorities had the dirt removed and the cistern filled with water, and it's been a tourist attraction ever since. Which is a good thing for the town, as twenty-first-century Ramla has been mostly known for its high crime and drug trafficking.

Fans of ancient underground cisterns might recognize a similarity to the Basilica Cistern in Istanbul, previously Constantinople. There has been some speculation that the design of the Pool of Arches was inspired by the Turkish cistern. Architecture aficionados will delight in knowing that the arches in the Ramla cistern are believed to be the first known pointed arches in the Arab world, a common design element of Islamic architecture. For instance, the Dome of the Rock, built in 691, had slightly pointed arches, but in the Ramla cistern we see for the first time a sharp point in the top-middle section of the arch. Moreover, the point spacing used at Ramla became the standard measurement for Islamic arches. Such characteristics can be seen wherever centuries-old Islamic structures were built, from Spain to Persia.

Today visitors can descend the modern metal staircase (built on top of the original stone staircase) and hop in a gondola to take a short cruise around the cistern. Think of it as an underground Venice in miniature. Or like you're floating around a flooded cathedral or an ancient mosque. And don't forget to look down to spot the carp and goldfish that live in the water.

Red and green lights illuminate the space, along with small openings in the ceiling where medieval residents of the city would drop buckets into the cistern to fetch water. It all helps create a moody ambience. An Arabic inscription on the wall claims the cistern was commissioned by Baghdad-based Caliph Harun al Rashid, a man who is the main figure in early editions of the *One Thousand and One Nights*, also known as the *Arabian Nights.*

Visitors tour the hidden waterways of the cistern by boat.

The Bell Cave in Beit Guvrin-Maresha National Park, Israel, a UNESCO World Heritage site.

BEIT GUVRIN-MARESHA
National Park

ISRAEL

━ ━ ━ ━ ━ ━ ━ ━ ━ ━ ━ ━ ━ ━ ━ ━ ━ ━

WHEN THE CITY OF MARESHA WAS UNEARTHED IN 1898, archaeologists were thrilled to discover an ancient town with walls encircling millennia-old houses and even two tall towers. But the real surprise was when they found more than two hundred underground rooms and passageways below the town.

The first mention of Maresha dates back to the Old Testament. In Joshua 15:44, it's listed as a town of the tribe of Judah along the western foothills of the Negev desert. It also later mentions that the Israelite king Rehoboam, son of Solomon, was responsible for building walls around it. The soft limestone that the town was built on, called nari, allowed residents to dig deep into the earth. They carved out not only wells for water but also spacious storerooms and halls. When the Kingdom of Judah eventually fell, Maresha came under control of the Edomites.

In the year 112 BC, Maresha suffered through more unrest as it was attacked by John Hyrcanus, a Hasmonean king, whose forces nearly razed the city. As the story goes, he converted the residents to Judaism in the process. And then seventy-two years later, in 40 BC, the Parthians came through, pounding the final nail into this once mighty city.

Another city known as Beit Guvrin later cropped up near the ashes of Maresha, and by the year AD 200, Maresha, now occupied by the Romans, was given the status of "city" once again. The Romans even gave it *Ius Italicum* (Italian Right) status, an

A remarkable feature of the underground spaces at Beit Guvrin are the honeycomb-like nooks that line many of its walls.

honor to non-Italian cities that treated them as if they were on Italian soil. Romans bestowed the city with the name Eleutheropolis, "city of free men."

Over the centuries, the area was quietly inhabited. From the Middle Ages on, Maresha was an Arab settlement—up until the 1948 Arab-Israeli War. After the conflict, a kibbutz was established near the ancient town, and then, in 1989, the national park was established on the site.

Today, tourists can visit Beit Guvrin-Maresha National Park and descend into some of the subterranean rooms that were occupied a millennium ago. There are extensive passageways below the earth where one can wander. There's an olive-oil press beneath one house. An aesthetically pleasing highlight is the Sidonian burial caves, where wealthy citizens buried their dead family members and decorated the walls with colorful scenes of flora and fauna, such as lions. Aboveground must-sees include Saint Anne's Church, a Byzantine-era sanctuary that medieval crusaders from Europe had rebuilt, and the second-century-AD Roman amphitheater, which was uncovered in the 1990s and had the capacity to seat thirty-five hundred spectators, who would have attended the amphitheater's many gladiator brawls. Amazingly, experts say only 10 percent of the area's underground caves have been fully excavated.

Unique to this site is a program managed by the Archaeological Seminars Institute called "Dig-for-a-Day," in which tourists and visitors can pay to be an archaeologist for a few hours, with the opportunity to make a new discovery while gently digging in a cave. In addition to feeling like an archaeologist for a while, one of the benefits of doing "Dig-for-a-Day" is getting access to caves that are normally off-limits to anyone but scholars.

An aesthetically pleasing highlight is the Sidonian burial caves, where wealthy citizens buried their dead family members and decorated the walls with colorful scenes of flora and fauna, such as lions.

And, of course, the other benefit is what you might find. So far several "tourist diggers" have unearthed objects that have changed historians' and archaeologists' views of the history of the area. A few years ago, two different tourist groups dug up pieces of an ancient plaque written in Greek. It turned out to be part of a royal proclamation by King Seleucus IV, who lived in the second century AD. The third and final piece of the puzzle was found in a private collection. After scholars put the three pieces together and translated the twenty-eight lines, a clearer picture of the famed Maccabean Revolt came into focus. The revolt, which is celebrated during Hanukkah every year, took place when Antiochus IV, brother of Seleucus IV, issued an edict forbidding many Jewish practices. The three pieces of the panel, now whole, gave a better idea of the run-up to the revolt and also provided historical context for the period. And if not for the "Dig-for-a-Day" program, it might not ever have been found.

But the most recent discovery in Maresha is not that of a marble plaque or even a new underground chamber. It's chicken bones. While archaeologists have found such bones in other ancient sites in the area, the finding in Maresha is significant because of how many were unearthed. After some more research, experts have come to a conclusion: the bones at Maresha are the first evidence of humans eating chicken meat.

Maresha has given the world more than just a better understanding of the ancient Holy Land. And with so much of it yet to be unearthed, there's going to be a lot more of this underground world to explore in the future.

Art featuring mythical creatures decorates the walls of the Columbarium Cave in Beit Guvrin-Maresha National Park, Israel.

The intricately engraved Column of Tears in the Basilica Cistern was carved over a millennium ago.

BASILICA CISTERN

ISTANBUL, TURKEY

WHEN THE BASILICA CISTERN WAS COMPLETED IN AD 532, its architects figured it would be filled up with water and that would be that: the ambient space (with twelve rows of twenty-eight Ionic and Corinthian columns each) would never be seen again, except by the occasional carp that managed to spill into the cistern from one of the aqueducts that feed into it. Fortunately, they were wrong.

Built during the reign of Emperor Justinian—who also is responsible for the nearby gargantuan sixth-century Christian church-turned-mosque, Hagia Sophia—the 105,000-square-foot Basilica Cistern was built by up to seven thousand slaves. It was originally underneath the now-lost-to-history Stoa Basilica (hence the name of the cistern). It may be just one of many underground water-filled spaces beneath the ancient city formerly known as Constantinople, but it also happens to be the most ambient and the largest, at 453 feet by 212 feet, making it one of the biggest Roman relics on the planet. If it were filled with water to maximum capacity, it would hold an astounding 17.5 million gallons. In ancient times, water flowed into the city from twelve miles away and was stored in cisterns like this one, which would then feed various public fountains around town, providing water for drinking and cleaning. Some historians believe the Basilica Cistern was the reservoir from which the emperors got their water. Cisterns were also important because when the city was under siege and the water supply from the aqueducts was cut off, reservoirs like this provided Constantinople's residents with water for months on end.

When the Ottomans took over the city in 1453, thus officially ending what was left of the Roman Empire and changing the name of Constantinople to Istanbul, they neglected the city's cisterns, including its biggest. They had a proclivity—as the Turks still do—for running water. Still waters were thought to harbor water-borne diseases, and so the Basilica Cistern became a trash dump (even a depository for dead bodies).

And then in 1545, one Petrus Gyllius, a French natural scientist and topographer who had been studying ancient Byzantine buildings and architecture, heard about the residents of central Istanbul fetching water (and sometimes fish) from deep holes in their cellars. The fishy thing about this fact was that the fish were of the freshwater variety, even though Istanbul is surrounded by salt-water. *Could this be the Basilica Cistern?* he wondered. So Petrus decided to do some exploring: with a rowboat, a sketchbook, and a torch, he paddled around the cistern. Afterward, he came to the conclusion that this was, in fact, the Basilica Cistern. But despite its "discovery," the cistern continued to function merely as a source for freshwater fish for the next couple of centuries. It was first restored in 1723 and then again in the early nineteenth century. Some cracks in the thirteen-foot-thick walls were repaired in 1968. In 1985, the city renovated the space once again, and in 1987, it opened to the public for the first time in centuries.

After the water was drained during a cleanup and restoration, there was something of a surprise: two sculptures of Medusa appeared, acting as plinths for two columns.

Twenty-first-century visitors can descend fifty-two stone steps into the earth below ancient Istanbul to feast their eyes on this stunning space, which sits almost thirty feet below the streets of the city. Suspended walkways allow visitors to explore the area, and the dim lights create a mysterious atmosphere. The Turks had their own name for the cistern: Yerebatan Sarayi, "sunken palace," because without water, that's truly what it resembles, as the 336 pillars present a majestic scene, almost as if visitors have stumbled upon a stone forest.

After the water was drained during a cleanup and restoration, there was something of a surprise: two sculptures of Medusa appeared, acting as plinths for two columns. One head is tilted sideways and one is completely upside down. Because the cistern is made up of recycled stone from ruined temples throughout the empire, the large medusa heads could be serving just as a foundation for a column. But, this being ancient Constantinople, there's a less practical, more mysterious explanation that some have given: the Christian emperor Justinian, building the Hagia Sophia nearby, took this pagan symbol of Medusa and buried it here sideways and upside down at the bottom of a pool of water, thus symbolically drowning the old religion.

Medusa heads serve as the base for some of the columns in the cistern. Some experts think they were used to achieve equal column height.

And still another common theory is that they are irregularly placed to negate the deadly Gorgon's gaze, by which a mortal would be turned to stone in the event of making eye contact with the serpentine-haired woman from Greek mythology. No one really knows why the medusa heads were placed here. Just don't look at them for too long!

The Column of Tears, a thirty-foot stalk of marble with intricately carved tear-shaped images etched into it, is said to bring luck. But not to just anyone. Those who may be granted good fortune from the column have to pass a test of skill first: place your thumb into a hole, located in the center of one of the teardrops, and spin your hand clockwise. If you can manage to turn your hand 360 degrees, congratulations: your wish may come true.

James Bond fans may recognize the cistern from the 1963 film *From Russia with Love*. More recently, the Basilica Cistern was used in the climax scene of the adaptation of Dan Brown's *Inferno*, in which the film's antagonist attempts to blow the cistern up. Spoiler alert: he fortunately fails.

Despite twenty-five massive earthquakes that have rocked the city, the Basilica Cistern and its 336 pillars are still standing strong, hardly looking a day over one thousand years old.

A stunning view of the Basilica
Cistern in Istanbul, Turkey.

Hot air balloons over
the Anatolia region
of Cappadocia.

DERINKUYU

CAPPADOCIA, TURKEY

I N 1963, A MAN IN CAPPADOCIA, TURKEY, WAS RENOVATING HIS HOUSE. Hoping to make a little more room in the basement, he decided to knock down a wall. As he began chipping away with the sledgehammer, he soon realized there was a lot more space behind that wall than he thought. In fact, there was a secret passageway and steps leading down into the darkness. Intrigued, he lit a torch and began a cautious descent. The stairway led him across a bridge and then to another incredible discovery: the ancient underground city of Derinkuyu.

The region of Cappadocia is recognizable for its many "fairy chimneys," the skyline of tall, twisting, squiggly volcanic rock formations that appear throughout the area. There could be as many as two hundred subterranean cities lying beneath them. The most captivating, though, is Derinkuyu, which is not the oldest nor the biggest, but it is the deepest, at eighteen floors, or 285 feet, below the ground. At the peak of the underground city's period of habitation—during late antiquity and the medieval periods—it was equipped with everything a human needed to live: washing facilities, toilets, wells, ventilation, storage areas, kitchens, and churches. Water for drinking and bathing was accessed via an underground spring. There was an olive oil press and a wine storage room—because if you're going to hole up underground for a while, olive oil and wine are necessary for survival. The whole network was big enough to house up to twenty thousand people.

The origins of underground cities in Cappadocia go back millions of years, to when a series of volcanic eruptions spread ash throughout the land. The ash eventually

"Because of these underground lairs, the region of Cappadocia became a safe haven for those persecuted in early Christianity, as converts could worship safely underground."

The mazelike passageways and tunnels in the deep underground city of ancient Derinkuyu.

morphed into a soft, porous volcanic tuff stone. Over thousands of years, wind and water erosion created a magical-looking landscape of odd rock formations, stone spires, and winding valleys that make the region seem straight out of a Hieronymus Bosch painting. The soft stone also made it easy to create underground passageways and rooms.

For security, the denizens of Derinkuyu carved circular stones five feet thick, weighing eleven hundred pounds, that could be rolled over entryways to block unwanted visitors. If someone did penetrate the underground lair, the inhabitants had a plan for that too: every floor could be sealed off from the others. In addition, the mazelike warren of narrow passageways made it impossible for intruders to find their way and forced groups to walk in easier-to-attack single file lines. Some long hallways led to dead ends, and others contained deep pitfalls where the intruders would plummet to their death. And if all else failed, there was a three-mile underground corridor to the Kaymakli, a nearby subterranean dwelling, where the residents could take refuge.

Scholars point to the Hittites as the possible first residents of underground Derinkuyu. In the fifteenth century BC, the Hittites were spread throughout the region, from the Black Sea to the Levant. In a twelfth-century-BC war with the Thracians, the Hittites may have first begun digging what is now known as Derinkuyu as a hideout from the invaders.

And yet other historians and archaeologists believe the Phrygians were the chief diggers of the underground city. Part of a larger Thracian tribe, the Phrygian people ruled the area until the sixth century BC, when they were overtaken by the Persians. Phrygians were known for their building prowess, and thus some scholars believe that only they could have excavated such a sophisticated subterranean world.

Lastly, there are the Persians. There's mention in ancient texts of the mythical Persian king Yima's proclivity for having underground cities dug to house his livestock and men, leading still other scholars to believe Persians once had such cities, including Derinkuyu.

The earliest known reference to the subterranean dwellings of Cappadocia comes from Greek philosopher Xenophon, who wrote around 370 BC of the people of this area living in underground rooms large enough to house entire families, their food, and even herds of animals.

Because of these underground lairs, the region of Cappadocia became a safe haven for those persecuted in early Christianity, because converts could worship safely underground. One of the most famous visitors to the area was the apostle Paul, who came through here in the first century. In fact, the region became such an important Christian stronghold that by the time the religion was legalized in the fourth century, the residents of Cappadocia were a powerful force in shaping church doctrine. Its underground cities were instrumental in this evolution.

Ventilation shafts 100 to 180 feet deep provide oxygen for tunnels and rooms ten stories beneath the surface.

Derinkuyu is not necessarily a staid underground museum, a monument to early human history long gone. Nearby underground dwellings were used as a refuge as recently as 1839 by Turks looking to escape from an invading Egyptian army. Today, visitors must descend 204 steps into the earth, hunched over most of the way down, to witness Derinkuyu with their own eyes. They see the underground city as it may have looked centuries ago, minus the bustle of residents, of course.

So far archaeologists have confirmed there are at least thirty-six cities underneath Cappadocia. Experts believe there are many more below the surface. In 2014, one of the largest subterranean cities was discovered below the Byzantine-era castle Nevşehir. Excavations are not yet complete, but initial studies estimate that it could be a third larger than Derinkuyu. Until the completion of the dig, we will wait for the inevitable discovery of the next underground hideaway in the area.

A cross carved from volcanic stone by Levon Arakelyan during his excavation of a network of tunnels and rooms below his home.

LEVON'S DIVINE
Underground

YEREVAN, ARMENIA

I T ALL BEGAN WITH POTATOES. ONE DAY IN 1985 IN YERAVAN, ARMENIA, Tosya Arakelyan asked her husband, Levon, if he'd kindly dig an underground cellar underneath their house to store potatoes. Levon, a construction worker and forty-four years old at the time, nodded and said he'd get right on it.

So, the next day, he grabbed some hand tools—a shovel, a pickax, a chisel—and began digging into the earth. He had dug about two and a half feet before he hit impenetrable basalt stone. So he started digging to the right of it, carving out the soft volcanic rock between the basalt. And then he hit another basalt stone. Most people at this point would have just given up. But Levon had other ideas. Using the basalt stone to his advantage, he kept digging, hitting basalt and carving out volcanic stone to the right. He ended up with a long staircase, each step formed by the basalt stones he kept hitting, and a small room for potato storage. Then he thought that he'd keep digging to create a wine cellar for him and his friends to sip vino in after work. Little did anyone know at the time, but Tosya's request would end up becoming a divine calling for Levon. He said he began hearing voices in his head, pushing him to keep digging and digging, chipping away at the dirt to create a subearthly paradise. Not a day went by that he didn't have a shovel in his hands, heaving dirt up to the surface. Some days he labored for up to eighteen hours. He said that the more he worked, the more power he got. He never felt tired. The hand tools wore out before he did.

If you'd asked Levon, he would have attributed his cave-digging work ethic to a trip to Russia that he took a few years before starting his now-famous project. It was

Levon Arakelyan constructing his Divine Undeground.

there that he had an inexplicable epiphany: he claimed to have been visited by a phantom who told him he would one day perform a miracle, that he would build a temple to God. The voice didn't give any other details. Shortly after he began digging the potato cellar, he knew this was it, and he had to do more than just dig one small underground room to store tubers. He had to carve out a physical miracle below his house.

After some months, Levon had dug seventy feet below the surface of the earth, carving out small rooms, stairways, and corridors. Many of the circular rooms have Ionic pillars and columns carved into them as relief sculptures. Some rooms are adorned with carved Christian crosses. And then, perhaps because of the voices in his head directing him, Levon began creating small shrines in this newly dug cave system. He said the dimensions of each room, each hallway, were dictated to him by God before he started carving them out. There was one room, in particular, that he felt had an especially spiritual vibe about it; its location was also determined by divine force,

He claimed to have been visited by a phantom who told him he would one day perform a miracle, that he would build a temple to God. The voice didn't give any other details.

and today it is filled with several small rock pyramids topped with single burning candles. He claimed that the inexplicable power he felt in the room could soothe an individual's suffering and could help prayers be answered. And so he continued to chip away at the rock, fulfilling the prophecy of the voice that had accompanied him from Russia to his Armenian cellar.

In 2008, twenty-three years after he first began, with the cave not as complete as he would have liked it, the digging ended and the divine voice in his head stopped permanently: Levon died at the age of sixty-seven. His hope was to keep digging for at least thirty more years, carving out seventy-four rooms (eighty in total). Alas, he didn't make it, but we have at least his six-room subterranean stunner. All in all, he went through twenty hammers and excavated 450 truckloads of earth.

His wife has turned the cave complex into a shrine to her late husband. Getting there isn't easy. From the capital, Yerevan, take a taxi out to the village of Avan-Arinj and then start asking anyone within earshot for "Levon's house." Someone will point you in the right direction, and soon enough you'll be standing in the middle of Levon's miraculous masterpiece. Or, as it's now officially called, "Levon's Divine Underground."

Once you arrive, walk down the eighty or so basalt stairs. At the end of the staircase, a large cross is carved into the wall. The rock was too impenetrable, so Levon carved a cross into it, saying that he could not move on until the cross was finished. When the cross was complete, he began chipping away at another wall and recommenced his divine mission. The first room one encounters is a large round space, with alcoves dug into the walls displaying various altars. The five other rooms each have niches etched into the walls. Candles inside each niche give the space a welcoming and pious feel. Back on the surface, don't miss the diminutive museum in the house, displaying the hand tools that Levon used to carve out his underground oasis.

With so many visitors, his wife now stores her potatoes aboveground.

The thirteen-story Vardzia cave monastery, carved into the side of Erusheti Mountain, Georgia.

GEORGIA

ARDZIA WAS A MASSIVE COMPLEX: SIX THOUSAND APARTMENTS. Thirteen floors. Twenty-five wine cellars. A meeting room, a reception center, and a pharmacy. There was even a throne room and a church. This isn't an eccentric urban apartment building. It's more like a human anthill. At least that's what the Vardzia cave monastery in Georgia looks like at first sight.

Started in the late twelfth century, the Vardzia cave complex in southern Georgia lies about forty miles from the town of Akhalsikhe. Like most underground lairs, Vardzia was constructed in the interest of self-preservation. Though the complex was begun by her successor, it was Queen Tamar, who ruled over a vast swath of land in the Caucasus that stretched from Azerbaijan in the south to Cherkessia in the north, who really put her stamp on the cave complex.

Tamar began her rule in 1184. And because of her age—some sources say she was just twenty-five years old when she took power—and possibly her gender, her reign saw a perpetual series of men trying to usurp her. No one succeeded. Contemporary Georgians revere this "warrior queen" for her strength and courage.

Fearing invading Mongols would wreak devastation upon the population, she ordered more caves be dug into the mountain. And so that the Mongols or any other outsiders could not penetrate the complex, the only way in and out was via a secret door on the embankment of the Kura River.

The cave complex was dug in a few different phases: chipping at the rock began during Giorgi III's rule in the mid-twelfth century. Toward the end of the century, now

Candles glow within the still-active church at Vardzia monastery.

during Tamar's reign, the Church of the Dormition was created. In the third phase, the complex was deepened and enlarged, and a system of running water was constructed in the early thirteenth century. At this point, there were six thousand dwellings in the cave city. There were also at least twenty-five different wine cellars in the complex, which makes sense, as contemporary Georgians claim wine was first cultivated in Georgia. The monks also planted gardens on the various terraces on the cliffside. Fortunately for them, the soil in this part of Georgia is extremely fertile. The gardens, along with the irrigation system they created, made the cave city self-sustainable.

Frescoes, such as this twelfth-century portrait of Jesus Christ, adorn the walls of Vardzia's Church of the Dormition.

In 1283, about one hundred years after the digging at the cliffside began, a massive earthquake did serious damage to the monastery, destroying about two-thirds of the structure. But this didn't stop the devoted and determined monks from staying and rebuilding, from persevering, like the generations and generations of troglodyte religious friars who would come after them. Then, in 1551, the Persians marched through the area and further destroyed Vardzia, killing all the resident monks at the time and leaving the monastery abandoned.

From a distance, you can see where the cliff side of the cave city caved in from the earthquake, exposing the honeycomb that existed inside the mountain.

The most important part of the Vardzia —at least for the monks who have resided here over the centuries—has been the Church of the Dormition. Carved into the rock, the spacious room measures twenty-seven by forty-five feet, with thirty-foot-high ceilings, and every inch of wall and ceiling space is plastered with stunning murals. Painted between the late twelfth and sixteenth centuries, the murals have had an indelible influence over the development and evolution of later Georgian mural painting. Art historians delight in the vaults of the upper walls, where the life of Christ—from the annunciation to the last supper to the crucifixion to the Virgin Mary's ascension (and everything in between)—is colorfully portrayed. On the north wall are depictions of former Georgian rulers, including Giorgi III and Tamar the Great; beneath the image of the great queen is the inscription "God grant her a long life." On the same wall is a portrait of Rati Surameli, who comes from the noble Georgian family that financially supported the monastery. The inscription next to his image reads, "Mother of God, accept the offering of your servant Rati."

Though Vardzla was abandoned in the mid-sixteenth century (after the Persian attack), monks eventually gravitated back to the cave city in the twentieth century and took up the troglodyte life once again.

Today the population of the Vardzia caves is limited to just seven monks whose raison d'être has been to help preserve the monastery and guard against its further deterioration. From a distance, you can see where the cliff side of the cave city caved in from the earthquake, exposing the honeycomb that existed inside the mountain. Here the monks still live like their brethren from past centuries, getting their drinking and bathing water from an ancient underground spring.

Visitors have access to about three hundred rooms and corridors. With many winding, twisting passageways ascending and descending, exploring Vardzia can be a dizzying journey, but there are very few sights in the world as unique.

SALT MINES

WIELICZKA SALT MINE, Poland

THE SALT CATHEDRAL OF ZIPAQUIRÁ, Colombia

SALINA TURDA, Transylvania, Romania

STRATACA, Hutchinson, Kansas, USA

T HERE'S SOMETHING intriguing about salt mines. Perhaps it's that they're burrowed so far into the earth. Or that it's easy to carve things into the walls. Or maybe it's just that we love the taste of salt so much. Whatever the case, sprinkled around the planet are a dash of salt mines that are worthy of our attention—they say something about the human need to make beauty out of our environment, even when it is hundreds of feet below the surface of the earth. There are countless examples, from Kansas to Kraków, Colombia to Romania.

In 1247 in the small Polish town of Wieliczka (pronounced *vay-leech-kah*), about ten miles southwest of Kraków, some peasants discovered salt in the ground. Suddenly, it seemed like everyone in the town began digging. And digging. And digging.

It may seem odd that the peasants acted like they'd struck gold when salt was discovered. But sodium chloride pretty much was considered white gold in the past. Roman soldiers were sometimes paid in salt, or *sal* in Latin, hence our word for "salary." In fact, salt was so important to the Roman Empire that one of their most popular roads was the Via Salaria, the Salt Road. The wealthy Republic of Ragusa (today Dubrovnik), sandwiched between the Ottoman Empire and the Venetian Republic, was able to fend off its powerful neighbors for centuries because it was extremely rich in salt.

Which is why back in Wieliczka, once a sizeable hole in the ground was dug, the Polish monarchy took over what would become a salt mine. Over time the Polish throne would earn a third of its income from Wieliczka and another nearby salt mine, Bochnia. Workers labored around the clock mining for salt at Wieliczka. So much so that the king began offering masses for the workers inside the mines. In 1896 miners began carving religious symbols into walls, creating four different chapels in the seven main chambers: there's a large relief sculpture of Da Vinci's *The Last Supper*, a life-size crucifixion scene, and, created more recently, a salt rock sculpture of Pope John Paul II. There are also secular salt rock sculptures, including intricate chandeliers and even the seven dwarves of Snow White fame.

The mine is one thousand feet deep in places, and the tunnels go on seemingly forever. They actually stretch 178 miles. In medieval times, workers had to be lowered down (and pulled up) via a sturdy rope. Visitors in the twenty-first century have to traipse down seven hundred wooden stairs. During the three-hour tour, tourists only cover about two

A chandelier lights the Chapel of Saint Kinga in the Wieliczka Salt Mine Museum.

miles of the mine, which is just a mere 1 percent of the actual underground space. In the King's Chapel, a cavernous space with an impressive amount of religious carvings, visitors can even attend mass: Sunday mornings at 8:00 a.m. More than a million people per year descend into the salt mine, and over the centuries some famous personalities have stopped by, including Pope John Paul II, Goethe, and Copernicus.

There have also been some rather wacky activities that have taken place below the ground here: in addition to weddings, parties, and conferences, people have bungee jumped inside the mine, and an adventure seeker attempted to windsurf in the mine's underground lake. Someone even inflated a hot air balloon in one of the one-hundred-foot-high chambers and floated around, the first time ever a hot air balloon has had a flight under the earth.

The mining halted at Wieliczka in 2007, but the force of tourism marches on: the salt mines are one of the most popular day trips from Kraków for visiting tourists. And for those who suffer from allergies, there is a clinic—or, as they call it, a "health resort"—set up deep within the mine where guests can get treatment for allergies effecting the respiratory system and even spend the night there.

A view inside the Wieliczka Salt Mine, one of the oldest salt mines in the world.

The Salt Cathedral of Zipaquirá.

AND IF IMAGERY AT Wieliczka doesn't appear pious enough, head over to central Colombia to gawk at the underground Salt Cathedral of Zipaquirá.

Located 650 feet below the surface of the earth, the Catedral de Sal is housed in a former salt mine. Starting in the early nineteenth century, the salt mine's workers labored nonstop to dig up salt. In the 1930s, some workers carved an altar into one of the salt walls, so they could pray for their safety before heading deep in the mine for work. In 1950, they began a more ambitious project: to carve an entire cathedral out of the underground salt space. Three years later it was done. Miners and nonminers came to pray at this twentieth-century architectural wonder for over four decades. And then in 1990, structural issues started to emerge, and the cathedral was closed for safety reasons. Five years later, they began making a new cathedral of salt, this time two hundred feet below the original. Over a quarter of a million tons of rock and salt were removed during construction. They say sequels rarely top the original. But in this case, the new cathedral was even more grandiose than the previous one. When it opened around the turn of the millennium, eighty-four hundred people attended the first mass, praying in front of the same sixteen-foot-high cross that had graced the altar of the first salt cathedral. After walking through a long tunnel, the ceiling illuminated with colorful hues, you emerge into an entryway bedecked with sculptures depicting the twelve Stations of the Cross, before entering the cathedral, the interior bathed in soft blue light, the plus-size cross at the altar dwarfing the rest of the room. The cathedral walls are etched with biblical scenes, including a recreation of the *Pietà* and a rendition of Michelangelo's *The Creation of Adam* from the Sistine Chapel.

IF YOU LIKE YOUR SALT MINES less devout and more diverting, head to Romania to the Salina Turda Salt Mines, located about 250 miles northwest of Bucharest. Here amusement park and historic salt mines merge into one. Descend 368 feet below in a slow-moving elevator, where miners have been excavating salt since the eleventh century. Before it was mined, the massive now-cavernous space once held three billion tons of salt. People have been digging up the mineral from Salina Turda since the days when this part of Europe was a Roman province called Dacia. Gone are the grueling days of pain and suffering from the terrible medieval working conditions; today the mine is known for its recreational activities. Visitors can play miniature golf, shoot some hoops on the basketball court, go bowling, play table tennis and billiards, and ride on a Ferris wheel, all of which sit on a platform above water at the bottom of the mine. You can also take out a rowboat on the water. And for those with respiratory problems, Salina Turda has a halotherapy center.

The salt mining stopped in 1932, and then the massive cavernous space was used to store cheese and other foodstuffs. During World War II it became a bomb shelter. The theme park aspect of the mine opened in 2010 and has been immensely popular: a half a million people visit Salina Turda every year.

No matter how odd it might seem to marry twenty-first-century recreation with medieval mines, there's no denying looking up at the 350-foot-deep conical hole you're standing in is an incredible feeling. It's worth it just for that.

IT'S ALSO WORTH JOURNEYING to Kansas to get a glance at what is going on underneath the surface. Long before Dorothy was wishing to transport herself there from Oz—in fact, 275 million years ago—the

The underground lake at the Salina Turda Salt
Mines now holds a modern-day amusement park.

The Ferris wheel at Salina Turda.

"Visitors can play miniature golf, shoot some hoops on the basketball court, go bowling, play table tennis and billiards, and ride on a Ferris wheel, all of which sit on a platform above water at the bottom of the mine."

The Kansas Underground Salt Museum, Strataca, offers tours that take visitors deep into its past, with displays on geology, nuclear waste storage, and, as shown above, railcars.

Permian Sea dried up over the middle of America. Salt was left in its wake. Lots of it. In fact, twenty-seven thousand square miles of it. But the epicenter (i.e., the saltiest section) is 650 feet below the earth in Hutchinson, Kansas. In 1887, Ben Blanchard was drilling on his land, hoping to strike oil. He found salt instead. And thus, the world's most popular seasoning became big business in this Kansas town.

Known today as the Kansas Underground Salt Museum, or Strataca, the vast caverns inside the mine—there are 980 acres in all—are supported by salt columns, some of them forty square feet in diameter. The Hutchinson Salt Company, mining here since 1923, still excavates about five hundred tons of the briny seasoning each year. Because of

the mine's consistent sixty-eight-degree temperature, it is ideal for storage. The company Underground Vaults and Storage keeps important documents here as well as the original master film prints of several important Hollywood movies, including *Gone with the Wind* and, fittingly enough, *The Wizard of Oz.*

There are fifteen salt mines in the United States, but only Strataca is accessible to visitors. Since 2007 people have been able spend a fun day (or entire night) sixty-five stories into the earth, partaking in various tours. A visit begins with an elevator descent into pitch-black darkness. Then guests can join a murder mystery dinner adventure—called "Murder in the Mine"—or ride the Salt Mine Express, a journey that takes you through the history of the mine, showing portions of the shaft that look exactly as they did a half a century ago. Thrill seekers might enjoy the "Dark Ride," a tram that pulls you through the most unlit parts of the mine, explaining mine hazards. For an even more extensive tour, the "Safari Shuttle" equips visitors with a hard hat and headlamp to descend into abandoned territories of the mine, where a 275-million-year-old crystal pod is one of the highlights.

Strataca is home to the Batman costume worn by George Clooney in *Batman & Robin*.

Across the
INDIAN OCEAN

An entrance to Biete Amanuel, one of nearly a dozen rock-hewn churches in Ethiopia.

BIETE AMANUEL

ETHIOPIA

I AM WEARY OF WRITING MORE ABOUT THESE BUILDINGS, BECAUSE IT seems to me that I shall not be believed," wrote Portuguese priest Francisco Alvares in the 1520s. "I swear by God, in whose power I am, that all I have written is the truth."

The buildings he was referring to are the rock-hewn, underground churches of Lalibela, which as a collection is one of the great wonders of the world, even if these jaw-dropping subterranean African structures get overshadowed by the Pyramids of Giza to the north. That they're not drawing hordes of tourists yet should give adventurers all the more reason to head for northern Ethiopia.

Not that Lalibela is the *only* wonder of this proud East African nation. There are over two hundred rock-hewn churches throughout Ethiopia. But the eleven churches in Lalibela, carved out of red volcanic stone and connected via underground passageways, are the most striking and most accessible. For members of the Ethiopian Orthodox Church, the Lalibela churches are a major pilgrimage site, right up there with the supposed Ark of the Covenant, kept in a church in Axum, 250 miles north of Lalibela. Yet only one hundred thousand non-Ethiopians come here per year, a pittance of a tourist population considering these subterrestrial churches are a historical and cultural treasure.

And among the eleven, the church of Biete Amanuel—the House of Emmanuel— might be the most stunning. Step down the steep staircase and into the superlative spiritual space, which was carved from one giant piece of earth. Before entering the church, visitors must remove their shoes, but have no fear: everyone is assigned a shoe porter who will look after them. Some of the eleven Lalibela churches are in a cross form, such as Biete Ghiorgis, the House of St. George. Some are carved into the face of a rock, while others are freestanding, like blocks carved forty feet deep into the ground. Some have simple facades, and others, such as Biete Amanuel, are marked by horizontal grooves etched into the stone and by cruciform windows.

Biete Amanuel's design is not random. It is, in fact, the best example of the twelfth-century Aksumite architecture. The Aksumite Empire ruled Ethiopia and Eritrea from the second to the tenth century AD. The revival of Aksumite architecture in the twelfth century, historians believe, reflects the then-ruling Zagwe dynasty's

The rock-cut churches of Lalibela, Ethiopia, were constructed by diggers who carved into the earth instead of building a place of worship on top of it.

desire to associate themselves with a previously powerful empire. Biete Amanuel's doors and windows, for example, are framed with wooden beams typical of the Aksumite design, offering just one example of how the rulers were attempting to visually link themselves to a once-potent past.

It's hard to conceive of now, but Lalibela was once a political center called Roha. The name was changed in the twelfth century to honor King Lalibela—who originally commissioned the churches—a few years after his death. King Lalibela is responsible for creating a sort of "new Jerusalem." When news arrived that Jerusalem had been taken over by Muslims, he wanted to create an alternative holy city. As European crusaders marched to the Holy Land, King Lalibela was having his men chip away at the volcanic rock to sculpt these magnificent churches.

Here you will see a peaceful account of local piety—people kissing the ancient walls, reading prayer books, and silently praying. The rock-chiseled churches of Lalibela are going strong after nearly a millennium.

The construction of the underground churches is a feat of medieval engineering. Instead of building from the bottom to the top, placing the last elements of a building at the highest point, the construction of Biete Amanuel (sometimes written as *Bete* or *Bet*) and the other subterranean, rock-hewn sanctuaries was done in the opposite manner: they were carved from the surface of the earth, with the last part being the base. And then there is the issue of flooding. The architects of the churches ingeniously invented a drainage system, complete with ditches and tunnels and even some pipes that shift water to baptismal pools. And some experts believe that the churches were constructed in an amazingly short twenty-three-year period.

One thing that makes Biete Amanuel and the other rock-carved churches so special is that they're still being used for their original purpose. Unlike, say, the Pyramids of Giza or Petra in Jordan, the locals continue to worship at these stunning structures, venerating and praying on a daily basis. In fact, if you want to see how the churches have been used for the last nine hundred years, wake up early and stop by Biete Amanuel or another church around five in the morning. Here you will see a peaceful account of local piety—people kissing the ancient walls, reading prayer books, and silently praying. The rock-chiseled churches of Lalibela are going strong after nearly a millennium.

Be sure to take some photos so that, unlike Francisco Alvares, the sixteenth-century Portuguese priest, people will believe you when you say you witnessed such a wonder of the world.

The centuries-old Ajanta cave system in India.

AJANTA CAVES

MAHARASHTRA, INDIA

OR OVER A THOUSAND YEARS IN THE EASTERN PART OF THE Maharashtra region in northwestern India, people strode past a horseshoe-shaped cliffside and didn't give a thought about the landscape, except to avoid the monkeys hanging out in the neem trees or perhaps to be on guard against a tiger crossing the path. And then one day in 1819, British officer John Smith, marching through the hinterlands of Maharashtra on a tiger hunt, came upon an incredible discovery: he noticed a cave opening midway up the cliff above the Wagurna River. He and his hunting party made their way into the opening. With a grass torch to light the dark space, they cautiously went inside, only to discover a cavernous hall bedecked with intricate relief sculptures of the Buddha etched into every part of the wall, from the floor to the ceiling. The centerpiece of the cave was a large statue of the deity, framed by a great stupa—a dome-shaped structure in a Buddhist shrine. The Ajanta Caves, long abandoned, had just been rediscovered.

Located 280 miles east of Mumbai, the city formerly known as Bombay, the thirty caves are now referred to by their assigned numbers. These are said to be the oldest known Buddhist-inspired caves and artwork in all of India. The grottos, dug out from a single 250-foot-high rock cliff of hard stone, were chiseled starting in the second century BC, though work continued until around the seventh century AD. And what's amazing about it all is that these were entirely a human construction—not simply a cave that was eroded by wind or water and then occupied by humans, but a man-made cave dug very deep into the rock and then lovingly but painstakingly bedecked with spiritual icons throughout, including soaring vaulted ceilings that could rival those of medieval European cathedrals.

After Buddhism's decline in India around the seventh century, the caves were abandoned as the population turned to Hinduism. As a result of the lack of human activity, the flora around the area grew and grew until the actual caves and temples were completely obscured, which is how they remained dormant until that fateful day in 1819 when John Smith just happened to catch a glimpse of a cave opening. After Smith came back from his hunt and reported what he'd stumbled upon, other curious Westerners gravitated to the caves, disguising themselves in local garb for fear some of the aggressive tribes in the area would have their heads.

The interior of the caves is a riot of engravings—lotus-position Buddhas and standing bodhisattvas of various sizes, colossal columns and stupendous stupas, griffons and other strange beasts, all carved out of one massive rock. One of the most jaw-dropping figures is the large reclining Buddha in cave 26, his elongated body stretching across the entire room. There are still-extant paintings, many of which tell

According to experts, the location of the caves— etched into a crescent-shaped cliff—is not random. In fact, they're positioned in a way to be congruent with specific celestial activity.

the story of the life (and reincarnations) of Siddhartha, the supreme Buddha. One of the most popular images in the caves is a painting of the bodhisattva Avalokiteshvara. A bodhisattva is a person who has been touched by the spirit of enlightenment and sometimes chooses not to go to Nirvana after death but opts for reincarnation instead, to continue to spread the word of the Buddha in another life. In the painting of Avalokiteshvara, the masterful artists beautifully capture the serenity on his face.

The two-thousand-year-old interior artwork may be stunning in its ancient beauty, but what had perplexed archaeologists and historians for a long time was how the caves, some of which are etched into the cliffs seventy feet above the ground, were actually made. Of course, determination and strength played an important role. But it turns out that the stone carvers and laborers hung from ropes, suspended some one hundred feet from the top of the cliff, and chipped away the stone, creating the beginnings of a cave opening.

According to experts, the location of the caves—etched into a crescent-shaped cliff—is not random. In fact, they're positioned in a way to be congruent with specific celestial activity. Some caves look east, some look west, which means grottos are positioned to catch the sunrise and others, the sunset. But their positioning is even more sophisticated than that. Take cave 19, for example. It was chiseled in its spot to perfectly catch the rays of the sun on the winter solstice. When the sun's light hits

A reclining Buddha in cave 26.

the back of the cave, a relief sculpture of Siddhartha Gautama, the founder of Buddhism, glows. On the opposite end of the spectrum, cave 26 performs a similar function but on the summer solstice. That they got the precise measurements for this to work so perfectly on both solstices reveals a sophisticated knowledge of mathematics, the stars, and planetary alignments.

During the centuries that the caves were in use, they were inhabited by monks, particularly during the monsoon season, and accommodated travelers who were passing through the area.

Back in the early nineteenth century, when John Smith had accidentally rediscovered the Ajanta Caves, he eventually resumed his tiger hunt. But not before carving his name and the date, April 1819, into the relief sculpture of a bodhisattva. Since then, visitors have carved their names or the names of their lovers into the walls of the cave. And with the exception of the deteriorated wall paintings and the stolen heads of five Buddhas, the Ajanta Caves remain largely preserved, despite the massive clutch of high-season tourists who traipse through the caves in complete wonderment every year.

Ajanta is made up of thirty caves, including beautiful artwork and shrines.

The Temple of the Great Kings at the
Dambulla Cave Temple, Sri Lanka.

DAMBULLA CAVE Temple

SRI LANKA

FROM TWELVE MILES OUT, IT LOOKS LIKE A SPACE ODDITY, SOME frighteningly large stone object that could have fallen from another galaxy, landing smack in the center of Sri Lanka. The Dambulla Cave Temple, also known as the Golden Temple of Dambulla, is carved into a plus-size granite rock surrounded by lush green forest, located about forty-five miles north of the town of Kandy. In fact, Dambulla is at the center of a historically important triangle: the three precolonial capitals, Kandy, Polonnoruwa, and Anuradhapura. Towering 525 feet above the ground, Dambulla is the best preserved temple of its kind in Sri Lanka. From certain points, the rock offers incredible views of the verdant jungle; you can even see the magnificent rock fortress, Sigiriya, resting in the distance.

The rock of Dambulla is so overwhelming, it almost belies what a visitor sees on the inside of it. There are 157 Buddha statues, all either standing or seated in the lotus meditative position. There are three statues of Sri Lankan kings, four sculptures of Hindu goddesses and gods, and about twenty-two thousand square feet of intricate and eye-dazzling murals on the wall and ceiling.

The caves actually existed before the complex became a Buddhist spiritual center. In fact, archaeologists have found human remains near the rock in which the caves are hewn that are up to twenty-seven hundred years old, leaving many experts to conclude the underground site was used for much longer than originally thought. But the origin story of the Dambulla cave complex goes back to the third century BC. That's when, magically, a bamboo tree sprouted up here at the exact moment King

In stark contrast to its more colorful interior, the extended white veranda of the Dambulla cave entrance is a welcome sight for visitors who hike to the temple.

Devanampiya Tissa converted to Buddhism, thus forever connecting the caves to both the supernatural and Buddhism.

In the first century BC, King Valagamba of Anuradhapura, also known as Vattagamani Abhaya, used the caves as a hideout because of invasions from southern India. He stayed in the caves for fourteen years before reestablishing his kingdom. After his return to Anuradhapura, he created a monastery at the grottos, about the same time that many Buddhists began using caves all over the island as a spiritual retreat, a haven for meditation and spiritual contemplation. As the decades and centuries went on, other kings added to the cave temples, commissioning murals here and new statues there. The caves, still relatively austere, got something of a makeover by the twelfth-century King Nissankamalla, who had the many Buddhist statues gilded and additional murals painted on the walls and ceiling. Henceforth, the caves

The origin story of the Dambulla cave complex goes back to the third century BC. That's when, magically, a bamboo tree sprouted up here at the exact moment King Devanampiya Tissa converted to Buddhism.

became known as Swarna Giriguhara, or Golden Rock Caves. After this, the Dambulla Cave Temple was known throughout Sri Lanka as the chief spiritual center of the island.

For modern-day visitors, getting into the caves is interesting in and of itself. Go through the modern kitschy structure, topped by a one-hundred-foot-tall seated golden Buddha; it has been claimed this is the largest Buddha statue on the planet, but there are others in Sri Lanka alone that are bigger. Visitors and pilgrims alike must ascend barefoot up a sloping path and the stairs that follow. From here, there's a fantastic view over the surrounding forest. Then step inside the complex, which consists of five separate caves of various sizes and extravagance.

You'll immediately enter the Cave of the Divine King. Meet the forty-six-foot-tall Buddha statue, chiseled out of the rock, which dominates the space. Then there's the Cave of the Great Kings, the grandest of all the cave chambers, featuring fifty-six Buddha statues, as well as a sculpture of the King Vattagamani Abhaya himself and the twelfth-century King Nissankamalla. The chamber is quite large, measuring 170 feet long and 75 feet wide, with a ceiling 20 feet high. Keep an eye out for the spring leaking water through a crack in the ceiling; tradition says that the water has healing potential. The cave ceiling also is covered in murals depicting the life of the Buddha, many of which are painted in vibrant red and yellow hues. The third cave, known as Great New Monastery, is clad in paintings from the eighteenth century and features fifty Buddha statues. In cave four, known as Western Temple, seated Buddhas

An unexpected detail—colorfully painted soles—on the feet of a reclining Buddha in the Golden Temple at Dambulla.

highlight the aesthetics of the room. The fifth cave, the smallest of the rooms, is dominated by a thirty-foot-long reclining Buddha, the position of Siddhartha Gautama on his death bed. A statue of the Buddha's loyal servant, Ananda, is situated nearby. Several other life-size Buddha sculptures are sprinkled around the room.

On your way out, expect to be inundated with the cottage industry of souvenir salespersons, as well as attention-seeking monkeys.

MODERN LIFE
Underground

MATERA AND CALCATA, Italy

COOBER PEDY, Australia

MATMATA, Tunisia

L IVING "UNDERGROUND" has long been a metaphor for going off the grid. In Dostoyevsky's *Notes from the Underground*, it is a symbol of humans living in complete isolation, a rejection of society. But sometimes we *choose* to literally go underground: to dwell below the surface of the planet because life is more soothing down there. From Italy to Australia to North Africa, humans have made homes for themselves by burrowing.

Matera may look at first like a typical southern Italian town: there are narrow cobbled lanes, intimate piazzas lined by coffee bars and salt-of-the-earth restaurants, and a ramshackle collection of stone houses perched on a hill that's crowned by a church steeple. But walk down to the area called Sassi, the Italian word for "stones," and you'll see it is not ordinary at all: Matera is dotted with up to fifteen hundred caves, many of which have been inhabited since the Paleolithic era. During the Renaissance, residents created vaulted archways inside the caves and adorned the facades with elaborate stonework.

Matera, once known as the "shame of Italy" because of its dire poverty, was "renovated" in the 1950s. The government moved the poor inhabitants out of the caves of Sassi and placed them in public housing projects. The caves sat abandoned for decades, save for a pack of wolves that took up residence in

One of the bell-shaped former cisterns that have been turned into a room at the Corte San Pietro hotel in Matera.

them. And shabby Matera took on the role as a stand-in for historic Jerusalem in films such as *The Passion of the Christ* by Mel Gibson and *The Gospel According to St. Matthew* by Pier Paolo Pasolini.

In the past decade, residents began renovating the caves of Sassi into private homes. Now there are even cave restaurants, cave cafés, cave clubs, and cave boutiques. Take, for example, Corte San Pietro, a five-cave hotel stylishly bedecked with flat-screen TVs, plush armchairs, and a minibar. And, of course, there's Wi-Fi.

ABOUT TWENTY-NINE MILES north of Rome is Calcata, a village built above a virtual honeycomb of caves. The town's stone houses are dug from the same volcanic tuff rock on which they sit, making it look like they magically sprang from the earth. For centuries, the residents of this town used the caves to store perishable food and wine. But in the 1960s, the village was condemned for fear its cliffs were crumbling. As the residents moved out, heading to a newly built nearby village of Calcata Nuova (New Calcata), hippies and artists moved into the medieval *borgo*, ascending through the hill town's gate (the only entrance and exit). They bought the homes for a few lira and moved in, patching up the cobblestone lanes and giving the exterior of the houses a new coat of paint. And those caves? They put them to good use. The grottos became residences, art galleries, and restaurants. Longtime Italian resident and American-born mosaic artist and chef Pancho Garrison managed to carve a spacious restaurant, La Grotta dei Germogli, from one of the caves. Then there's artist Giancarlo Croce's studio and exhibition space, Studio d'Arte Porta Segreta, where a long staircase descends into the middle of the tuff mountain.

And if you're not caved out in Calcata yet, head down to the base of the rock on which the village sits. Carved into the rock are caves created by the Faliscans, an ancient pre-Roman people who existed in this part of central Italy. The Romans eventually conquered and wiped out the Faliscans in the third century AD. They destroyed nearly every trace of the Faliscan culture along the way. But because of the Roman aversion to all things dead, they left these caves relatively untouched, since they were tombs. Today you can walk below Calcata in a verdant regional park and explore these caves, the (empty) stone coffins still in place.

BUT ITALIANS ARE NOT the only people on the planet with a proclivity for cave dwelling or living under the earth. Australia's Coober Pedy, the opal-mining capital of the world, gives new meaning to the term "down under."

Drive through this town of thirty-five hundred people, about 525 miles north of Adelaide, and you'll only see a smattering of buildings: a few hotels, shops, and a golf course. It's not a ghost town. It's just that an estimated 60 percent of the population of Coober Pedy lives about twenty feet beneath the surface.

In 1915, fifteen-year-old gold miner William Hutchinson wandered away from his work camp to look for water. But before he could quench his thirst, he found something else: opal. Lots of it. A year later, the "opal rush" was on, as people from as far as Europe moved to the new town of Coober Pedy to get in on the action. But they soon discovered that the hardest part about living there is the heat. Average summer temperatures, which hold steady at around 104 degrees Fahrenheit (and can reach up to 125 degrees), still kill about forty people a year. So the new residents began digging. They dug homes into

The dining room of Faye Nayler, a resident of Coober Pedy, Australia, who dug the home with the help of two friends.

cliffsides and mountains, etching out kitchens, living rooms, and bedrooms. In fact, the name Coober Pedy comes from the Aboriginal word *kupa-piti*, which literally means "white man's hole."

Today there are underground hotels, cafés, restaurants, and even two churches. There's an underground pub called the Desert Cave. And then there are fifteen hundred underground homes that locals call "dugouts." And that opal? Seventy percent of the world's opal still comes from Coober Pedy, mined by its cave-dwelling population.

In fact, when newer underground homes are being dug, one perk is discovering the precious stone. One hotel unearthed $360,000 worth of opals while digging its guest rooms.

The most well-known fictional resident of the underground homes in Matmata may be Luke Skywalker from *Star Wars*.

THE HEAT MAY ALSO BE ONE of the reasons why there are underground homes in southern Tunisia. Located in the small town of Matmata—three hundred miles from the capital, Tunis—the one hundred plus underground homes were constructed by first digging a large twenty-foot-deep by thirty-foot-wide pit and then carving out rooms into the sandstone. It's a custom that has been practiced here for at least two millennia.

Tales about how and why people began digging are legion. One of the most oft-told stories dates back to the Punic Wars (264–146 BC): Romans dispatched a group of Egyptians to the area to wipe out the local population and then establish Roman rule in the region. The locals fled farther south and then, just for good measure, dug homes deep into the ground to hide. At night, they'd emerge from their netherworld homes to attack the Egyptians, eventually sending them packing. This may be true, but if you pay a visit to southern Tunisia, where the average summertime temperature is ninety-two degrees, it would be easy to figure out why living underground would be much more comfortable. After all, the temperature in the cave homes hovers around sixty-eight degrees throughout the year.

The circular pits are more dynamic than they appear. Dug out into the circumference of the pit are three or four homes. The pit itself, then, becomes a courtyard, some of which have wells in the center. Inside the subterrestrial homes are several rooms complete with modern-day luxuries like running water, electricity, and Internet connections. Some of the pits are connected via underground passageways.

And amazingly, the people of Matmata remained generally closed off from the outside world for centuries. It wasn't until 1967, when there were twenty-two consecutive days of non-stop rain, and floodwaters damaged several of the subterranean dwellings, that the people of Matmata climbed out and pleaded for help. A delegation from the Tunisian government traveled to the area and were amazed to discover the troglodyte shelters. The Tunisian government then built an aboveground city for the residents, but most of them had their underground homes repaired and have continued living there to this day.

For visitors with an overwhelming urge to spend the night in one of these underground dwellings, the Hotel Marhala is the place for you: the basic rooms are comfortable and cool, away from the heat of the desert, and the lobby even has fast Wi-Fi.

For fans of *Star Wars*, visiting Matmata becomes a pilgrimage: one of the underground dwellings once doubled as Luke Skywalker's place of residence. Recent reconnaissance reveals he has since moved.

From
BEIJING
to the
STRAIT OF
MALACCA

A painted grotto in the Mogao Caves.

MOGAO CAVES

DUNHUANG, CHINA

ROUND SEVENTEEN HUNDRED YEARS AGO, A BUDDHIST MONK named Le Zun had a vision: he saw a thousand Buddhas, all bathed in a golden glow in a vast desert. The story, apparently recorded in an ancient book called *An Account of Buddhist Shrines*, foreshadowed what actually happened.

Located in northwestern China, the town of Dunhuang is a city of nearly two hundred thousand people in the province of Gansu. Today it may seem like something of a backwater, as Chinese metropolises Beijing and Shanghai get the lion's share of attention, but two thousand years ago, Dunhuang was smack in the center of the famed Silk Road. The city saw ample amounts of trade leave China and move west toward Europe. It also saw its fair share of foreigners traipsing through, as well as foreign goods that would turn up from time to time.

These days, though, Dunhuang is relatively off the tourist radar. So why would anyone today, save for aficionados of Silk Road history, come all the way to Dunhuang, a three-hour flight from Beijing? Perhaps we have Le Zun to thank for that.

Set on the Dachuan River, the Mogao Caves—fifteen miles southeast of Dunhuang—are carved into a cliffside and contain an astounding 735 caves, housing almost five hundred thousand square feet of wall and ceiling paintings and boasting over two thousand statues. Today it is one of the largest collections of Buddhist art on the planet. First built in AD 366, the grottos—also called the Thousand Buddha Caves—contain some of the finest Buddhist art in China, all of which was created over a thousand-year period lasting until the fourteenth century. The caves themselves are a product of the Silk Road. Pre-Buddhist China favored temples made of wood. But in the fourth century, Buddhists arrived from India, where—thanks to the heat—they had a proclivity for turning cool caves into temples, and the grottos at Mogao made for a perfect spot in which to create a temple to the Buddha.

There were more than one thousand caves, but some have been lost to history or not yet rediscovered. The mile-long cliffside in which the caves are carved contains

An entrance to the Mogao Caves, also known as the Caves of the Thousand Buddhas, in China.

more baroque interiors to the south while the caves in the northern section are relatively unadorned. Murals within depict the life of the Buddha as well as images of traders on the Silk Road. Artists gravitated to Dunhuang to help paint the murals; they lived in the northern grottos while working in the southern caves, often laboring all day to decorate the grottos. There are sculptures of bodhisattvas, warriors, and Buddhas in various positions—from standing to sitting to reclining.

The artwork in the caves reveals the extent of the riches the Silk Road brought to Dunhuang. But then in the fourteenth century, trading on the Silk Road dried up, and the cave complex was abandoned. Sand from the Gobi Desert eventually filled in some of the grottos and blocked the entrances to others. Thus, they sat empty for nearly five hundred years. In 1900, though, a Daoist monk named Wang Yuanlu unearthed a cave via a wall that had fallen down. Inside the room were fifty thousand documents, works of art, and tapestries dating back to the eleventh century. News of the discovery spread around the world, and the Mogao Caves experienced a resurgence of interest. Soon scholars and the generally curious were turning up from all corners of the globe. They were buying the manuscripts and taking them back home. Among the most notable finds was the Diamond Sutra,

The Mogao Caves—fifteen miles southeast of Dunhuang—are carved into a cliffside and contain an astounding 735 caves, housing almost five hundred thousand square feet of wall and ceiling paintings and boasting over two thousand statues.

a ninth century Buddhist text that many scholars believe is the oldest printed book in the world. Today, the documents can be found spread out around the planet, everywhere from Beijing to Berlin, Paris to Saint Petersburg. The manuscripts reflect the diversity of Dunhuang centuries ago. Thanks to the Silk Road, everything from Daoist to Confucian to Christian texts were found in the caves, written in several languages: Tibetan, Hebrew, Sanskrit, and Turkish.

Since the discovery of the manuscripts, archaeologists and art historians have helped restore at least 492 of the caves. And historians have traced the evolving styles and influences from the paintings and sculptures. The fourth-century creations, for example, reflect influence from Greece and India—evidence that not only physical objects were being traded and shared along the Silk Road, but ideas as well—while the works created from the seventh to the tenth centuries are done in the style of the Tang dynasty in China.

Many of the caves are accessible to visitors today. They just don't allow any photos, so a postcard of one of the many Buddha statues will have to suffice.

Three of the
eight thousand
terra-cotta warriors
unearthed in Xi'an,
China.

The Museum of
QIN TERRA-COTTA
Warriors and Horses

XI'AN, CHINA

IN 1974, YANG ZHIFA AND HIS FIVE BROTHERS WERE DIGGING A WELL in the Shaanxi province near the town of Xi'an. Instead of water, though, they dug up half of an ancient terra-cotta soldier and some other pieces of baked clay. Without much thought about if possession of such things could conjure the spirits or other unsettling forces, they began selling them to nearby cultural and historical centers to make some extra money. This soon piqued archaeologists' interest, and by 1976, a dig was underway. The team of archaeologists unearthed a literal army, which was much grander than anyone could have imagined: 8,000 soldiers, 750 horses, and 130 chariots. The army was part of what became known as the terra-cotta warriors, life-size terra-cotta Chinese soldiers (and their equine partners) designed and constructed to help protect the tomb of third-century-BC emperor Qin Shi Huang. The discovery would be one of the most monumental of the twentieth century.

The warriors are just part of the massive underground complex, which has not been completely excavated yet, including the lavish tomb of Emperor Qin Shi Huang. The tomb is housed within a 250-foot-tall pyramid. The humongous compound is meant to be a small-scale model of the city of Xianyang, the Qin capital: it contains both inner and outer "cities," stretching one and a half miles wide by nearly four miles long. The warriors, all of which have not been unearthed yet, were simply a garrison protecting the mausoleum.

The complex seems fitting for the man who would be crowned king at the age of

Though a significant amount of the terra-cotta warriors site has been excavated, there is still more to be uncovered.

> "The team of archaeologists unearthed a literal army, which was much grander than anyone could have imagined: 8,000 soldiers, 750 horses, and 130 chariots. The army was part of what became known as the terra-cotta warriors."

18

thirteen. He went on to unify the entirety of China in the third century BC. And work actually began on the massive mausoleum in 246 BC, thirty-six years before the emperor would be laid to rest.

Each of the terra-cotta warriors has different facial features.

Amazingly, just five years after the initial discovery of the warriors, a below-ground exhibition space on the grounds opened up to the public in 1979. What was an off-the-radar swath of northwestern China became one of the country's biggest tourist draws. With much of the area still being excavated, the site continues popping up in the news: there are now six hundred pits in the ground around the complex, as archaeologists continue to find significant relics of the past. The entire dig spans twenty-two square miles. And why hasn't the emperor's tomb been dug out yet? Legend suggests that it is booby-trapped with deadly mercury or arrows that will be released on those who enter.

Why hasn't the emperor's tomb been dug out yet? Legend suggests that it is booby-trapped with deadly mercury or arrows that will be released on those who enter.

And while not every excavation pit is accessible to tourists, there are three open to the public. They make up the Museum of Qin Terra-Cotta Warriors and Horses: the first pit, the spot where the farmers originally found the warriors, exhibits six thousand warriors, each one unique from the next. The 172,000-square-foot space is sixteen feet below the earth. The second pit, about 400 feet long by 320 feet wide by 16 feet deep, demonstrates how the terra-cotta soldiers looked when they were first found in the mid-1970s, as some of the statues lie broken while others stand upright and some are still half-buried in the dirt. There are over thirteen hundred soldiers in pit two, along with eighty war chariots. Pit three, the smallest at sixty feet by seventy feet, and perhaps the least impressive of the trio, exhibits warriors, many of them headless, and a smattering of terra-cotta horses.

In 2016, *National Geographic* reported on a wild new theory about how the terra-cotta statues were made: that Hellenistic artists—from Greece or within Greece's cultural sphere of influence—came to Xi'an to teach local artisans how to sculpt such figures. European DNA has been found in some of the bones recovered from the site, so it's possible. This radical idea, if true, would place Europeans in China fifteen hundred years before Marco Polo came marching through.

And so what ever happened to the five well-digging brothers who discovered the terra-cotta warriors? Some say they were cursed. Their land was taken over by the government to dig further. Their houses were destroyed in order to build structures for the future onslaught of mass tourism. One brother committed suicide. Two others died two decades after the discovery—they were both without jobs and penniless. The remaining brothers today earn about two dollars per day working in a souvenir shop. The world benefited from their find, but the brothers did not.

The immense G-Can Tunnels outside Tokyo, Japan, are also known as the Underground Temple.

G-CAN TUNNELS

TOKYO, JAPAN

HERE ARE MANY SUPERLATIVES WHEN IT COMES TO JAPAN: THE country is home to the oldest wooden building in the world. Tokyo is the most populous city on the planet. And the Sky Tree in Tokyo is the world's tallest tower.

And so on the outskirts of Tokyo, lurking beneath a grassy field, is another Japanese superlative: the world's largest sewer system—albeit, one that flushes out water, in general, and not human waste–tainted water. The Metropolitan Area Outer Underground Discharge Channel, or *Shutoken Gaikaku Hosuiro* in Japanese, is known by non-Japanese speakers as G-Cans for short. The impressive system is located just twenty miles from the center of Tokyo. Open up the door to a nondescript building in the middle of a field and descend a long set of stairs. Eventually you'll be deposited into a massive tank ringed by a forest of fifty-nine thick concrete pillars. The columns themselves weigh five hundred tons each and are six feet wide and sixty feet tall. Five stories high and as long and wide as a football field, this water shaft—known as the Underground Temple—is the main depository for the water. But that's not all.

There's also a four-mile-long tunnel that runs underneath the ground connecting five different below-the-surface tanks. The massive shafts, which also collect flood-water, are 230 feet deep. And the whole thing is powered by four turbines, the same engines that run a 737 jet airliner. When operating, the engines have the astonishing capacity to drain an eighty-foot-deep basin in one second.

The nearly $3 billion project, which began in 1992 and finished seventeen years later, helps protect thirteen million people in a part of Tokyo that is prone to flooding. One of the unique geographical characteristics of Japan is that 75 percent of the land is mountainous. This means there are many rivers that flow downward to the sea. And in times of heavy rain, those rivers tend to overflow. If this swath of northern Tokyo gets at least twenty-one inches of rain within a continuous three-day period, then rivers of the area, particularly the Arakawa, overflow and could flood the region. And it's even worse when one of the regular massive storms hits the city. One famous storm in 1910 ruined 4.2 percent of Japan's GDP. Throughout the years, these storms have pounded the city, leaving some neighborhoods completely submerged underwater. After six deadly floods occurred throughout the 1980s, including one storm in 1991 that flooded over thirty thousand homes (and resulted in fifty-two deaths) in northern Tokyo, city officials realized they had to do something drastic. The solution was the G-Cans.

Here's how the ingenious system works: Each of the five silos, located about a mile from one another, is near a river. When one of the rivers overflows—often during the monsoon season, from the beginning of June to mid-July—the extra water runs into the tanks. It's then channeled through the underground tunnels, which are about 30 feet in diameter and located 160 feet underground, to the "Temple," where it sits until it is eventually drained. Four Boeing 737-jet-engine-powered turbines pump into six underground tunnels—each about the length of a typical train—that carry the water into the Edo River, where it then easily flows into the sea, causing no harm to anyone. The turbine engines have the capability of pumping up to two hundred tons of water per second. So is this pricey project worth it? In the first five years of its existence, it was used seventy times.

The main tank, the massive space with the pillars, has been featured on numerous international TV shows and movies, including on Dutch and Australian TV. So, perhaps for some visitors it is déjà vu.

> *Four Boeing 737-jet-engine-powered turbines pump into six underground tunnels—each about the length of a typical train—that carry the water into the Edo River, where it then easily flows into the sea, causing no harm to anyone.*

There are free ninety-minute tours—unfortunately only in Japanese—for those curious about wandering around in the world's largest sewer. Non-Japanese-speaking visitors must have a Japanese speaker accompany them to explain the safety procedures. Before descending into the tanks, guests are given an overview, including an explanation of how the system works and why G-Cans was necessary. The guide shows satellite photos of Tokyo and points out the various rivers that tend to frequently flood. And after a visit to the project control room with its twenty television monitors and a brief stop on the roof to get a nice view of the nearby Edo River (if you squint you can see the Tokyo Sky Tree in the distance), it's time to head down to the tanks and one of the most brilliant water discharge channels on the planet, which lies unassumingly beneath a skateboard park and a soccer field.

In times when the system is in use, the tours are obviously canceled.

Visitors to the G-Can tunnels must be accompanied by a Japanese speaker in case an emergency evacuation is announced.

A stunning view of Phraya Nakhon
cave and its royal pavilion.

PHRAYA NAKHON Cave

THAILAND

———————————————————————

LOCATED NEAR KHAO SAM ROI YOT NATIONAL PARK, ABOUT THIRTY miles south of Hua Hin in the Prachuap Khiri Khan province, the Phraya Nakhon cave is off the beaten path, but it's well worth the journey. At the beachside village of Bang Pu, hire a long-tail boat to take you down the coast to Laem Sala beach, where you can find the trail that leads to the limestone cave. Or just stroll along the hilly path for about thirty minutes. Now for the hard part: walk up fourteen hundred feet of uneven, rocky steps, flanked by tropical trees and plants. As you hike, you'll pass by a few shrines, complete with statues of tigers, then pause to take a breather and enjoy a nice view of the Gulf of Thailand. Keep an eye out for local wildlife, including ground lizards and owls in the trees. Since the area is known for its hungry mosquitos, remember to bring your bug spray. You'll be rewarded with the majestic sight of the Phraya Nakhon cave. In the first of three grottos, have a look at the map of the cave system and the information board. Then cross over a bridge—named Death Bridge because several wild animals have apparently plunged to their deaths from it—and enter the second cave, where sunlight cascades from a large hole in the ceiling. Beams of light, like divine rays from heaven, shine down on a late nineteenth-century pavilion reminiscent of some kind of peaceful Shangri-La. Where natural light reaches the floor, trees have grown up in the back of the cave, making their home among stalagmites while reaching toward the stalactites hanging from the cave's ceiling. Be prepared to feel dwarfed in the cavern: the spacious grotto is 160 feet high.

Nestled in this otherworldly place, the blue and gold pavilion, called the Kuha Kharuehat, was built for the visit of King Chulalongkorn the Great, also known as Rama V. It was actually built in 1890 by artisans in Bangkok and then transported in pieces to the region. Carrying the building materials up the verdant hill and into the cave, where the parts were assembled, must have been grueling. Since that time other Thai kings have visited the cave: Kings Rama V and Rama VII both carved their names into one of its walls. But today, the cave hosts adventurous travelers and entertaining langur monkeys, who scamper around, playing and generally behaving mischievously.

Beams of light, like divine rays from heaven, shine down on a late nineteenth-century pavilion reminiscent of some kind of peaceful Shangri-La.

The cave is reputedly named after Phraya Nakhon, a ruler who stumbled upon the grotto about two hundred years ago when his ship was wrecked just off the coast here. Some people believe the cave is named after another prominent person named Nakhon who lived in the seventeenth century—though little is known about him. We may never know the truth—only that Phraya Nakhon is one of the most stunning caves you'll lay your eyes on.

Try to time your visit to the cave so that you arrive around 11:00 a.m. That's when the sun shines through a hole in the top of the cave, spotlighting the pavilion and giving it a supernatural glow. And because of the position of the sun, January is the ideal time to go in the morning.

After visiting the cave and descending the mountain, make your way to a restaurant near the beach, a perfect way to celebrate the journey to one of Southeast Asia's most memorable caves.

Prahya Nakhon cave's paviion in Khao
Sam Roi Yot National Park, Thailand.

A statue of a Hindu god welcomes
visitors to the Batu Caves, Malaysia.

THE BATU CAVES of Selangor

SELANGOR, MALAYSIA

LOCATED JUST OUTSIDE OF KUALA LUMPUR, THE BATU CAVES CAN BE seen from afar: giant limestone mounds with a couple of tiny, dark openings. And then you get closer, and the whole site is mesmerizing, complete with a 140-foot statue of Murugan (the biggest in the world), the Hindu god of war, which was added in 2006.

The Batu Caves, named for the river that flows by this limestone formation, consist of three spacious grottos (with a few smaller caves tucked away).

Some geologists estimate that the caves are four hundred million years old. They were formed by consistent levels of slightly acidic rain washing through limestone rocks, creating caves and then stalactites and stalagmites and other odd rock formations. The indigenous Temuan people used the caves, and much later, Chinese settlers used the grottos to fetch guano, excrement from coastal birds that doubles as excellent fertilizer. But it wasn't until 1878 when news of these fantastic caves began to spread around the world. That's when William Temple Hornaday, the first director of the Bronx Zoo—famed for putting a Pygmy from the Congo in a cage with monkeys—boasted far and wide about the caves he'd visited while in Malaysia. During the colonial period, British residents of Kuala Lumpur would take day trips to go exploring the caves by candlelight.

But the caves really got a boost about a decade after William Temple Hornaday began publicizing them, when businessman K. Thamboosamy Pillay and others began promoting the caves as a Hindu place of worship. From this point onward, the cave gained a regular influx of visitors. There wasn't an overt historical connection to spirituality with the caves but instead just a general vibe that there was something special here, a celestial sense. Some have speculated that the Hindu attraction to the cave's rocky mountains was because they vaguely resembled a smaller version of the Himalayas. And soon enough a shrine was built here, one dedicated to Lord Muraga, who also goes by the name Lord Subramaniam. A bit later, they added a shrine to Ganesh, and these days there's a whole spectrum of shrines to various Hindu gods. The caves, about three hundred feet up the rock, are accessible by 272 stone steps, which replaced the rickety wooden steps originally built in 1920.

Before heading up the long stairway to the Temple Cave, take the path to the art gallery cave, a grotto crammed with brightly hued statues of Hindu gods. There are also dioramas and murals with varying degrees of psychedelic curiousness. Facing the mountain, go left to get a glance at the Ramayana Cave, passing by a fifty-foot-tall statue of Hanuman, the Hindu monkey-man god and one of the main figures in the

Thaipusam festival goers fill the main cavern at Batu.

168

epic Indian poem called the "Ramayana." Inside the cave, colorfully painted statues that line the irregularly shaped walls tell the story of Lord Rama, as detailed in the poem. Along with the various colors of light that bathe the bumpy walls, the plus-size stalactites and stalagmites and winding stone staircases help to create a magical ambience inside the cave. One of the highlights of the cave is the sleeping giant Hanuman, peacefully resting while life-size servants and even an elephant stand around him.

After taking in the Ramayana Cave, make your way up the stone steps. About two-thirds of the way up, you have the option to explore the Dark Caves, where seven different chambers boast crazy rock formations and a whole set of plants, animals, and insects that only exist right here, including spiders, monkeys, and two different kinds of bat. There are one and a half miles of caverns and hallways that can be explored in the Dark Caves. For that reason, any brave visitor who wants to journey into them must be accompanied by a guide.

> *Along with the various colors of light that bathe the bumpy walls, the plus-size stalactites and stalagmites and winding stone staircases help to create a magical ambience inside the cave.*

Back on the main steps, reach the top to see the stupefying Subramaniam Swamy Temple, located in a massive cavern with three-hundred-foot-high ceilings. The interior of the cave has a glowing quality thanks to the rays of sunlight penetrating the grotto through holes leading to the exterior. The cave walls are lined with statues, some representing the six lives of Lord Subramaniam. If you want to explore deeper into this part of the cave, employees will dot your forehead with a red circle, thus allowing access to the parts of the temple grotto.

For those who want to see the caves in their full, maximum usage, go during Thaipusam, one of the most significant festivals in Malaysian Hinduism, held during the full moon between mid-January and mid-February (depending on the lunar calendar). Hundreds of thousands of Hindus and non-Hindus flock to Batu during the festival to watch devotees ascend the steps to the cave carrying *kavadis*, or "burdens," such as wooden frames connected to the wearer with hooks or a milk jug affixed to the top of the head with spikes digging into the flesh. The message: you cannot reach the divine without suffering and hard work.

One benefit of visiting at a nonfestival time is there may be more of a chance visitors can interact with the many monkeys that live in and around the caves. And "interacting," in this case, means monkeys stealing tourists' snacks, phones, and sunglasses.

The series of statues in the Ramayana
cave detail the eponymous Indian tale.

An employee of the Cu Chi tunnels museum dresses in Viet Cong garb and demonstrates the underground complex's camouflaged entrances.

CU CHI
Tunnels

VIETNAM

LOCATED FORTY-FIVE MILES FROM HO CHI MINH CITY, THE Vietnamese metropolis most locals still call Saigon, the Cu Chi tunnels are one of the major tourist draws of the area. They were also one of the reasons why the North Vietnamese defeated the Americans in the war that lasted from the mid-1950s to 1975.

The tunnels were extensive: at one time there were 120 miles of them, veining throughout the land from Saigon to the Cambodian border, allowing the Viet Cong, the North Vietnamese soldiers, safe passage throughout the area. For over twenty years, starting during the French occupation of the Southeast Asian nation, ordinary Vietnamese would dig their way underground in order to create this hidden network of passageways. The anthill-like tunnels not only acted as a way to sneak up on foreign troops, but also became "home" for many. During periods of American carpet-bombing, the inhabitants of the tunnels would remain underground for days on end. Some Vietnamese would come out of the tunnels only at night to scavenge for food or to work their fields. Health conditions were awful, with half of the denizens contracting malaria and almost everyone suffering from parasites. They shared the space with spiders, snakes, scorpions, rats, and other rodents and pests. There are a few different levels to the tunnels, the deepest about twenty feet below the surface and the shallowest about five feet below. The tunnels couldn't be too deep, as they needed ventilation. And yet, there were entire villages living under the ground, all for protection: to stay out of sight of the enemy and to avoid aerial bombing.

Despite the conditions inside the tunnels, the subterranean passageways were an ingenious way for the Viet Cong to counteract the technological superiority of the Americans. They might not have had the latest war tech, but they had a vast network of underground tunnels that would allow them to surprise the enemy; this was part of the home field advantage. Also, the tunnels were tiny, often just four feet high—too small for American soldiers to navigate easily. They twisted and turned, snaking through the ground in South Vietnam, some of them even booby-trapped—with trapdoors that would send a soldier plummeting to the bottom of a pit filled with life-ending spikes. This discouraged any "tunnel rats," American soldiers who would descend into the tunnels with only a gun, a knife, a flashlight, and a long string (to find their way out), hoping to force some Viet Cong up to the surface.

These subterranean spaces were the birthplace of the notorious Tet Offensive. This series of surprise attacks by the Viet Cong against southern Vietnamese and American troops on January 30, 1968, was one of the largest military campaigns of the war.

The Americans stepped up their game to try to get the North Vietnamese out of the tunnels. Operation Crimp, in January 1966, led by American and Australian forces, attempted to bomb the underground Viet Cong headquarters. The project was only partially successful. A year later, Operation Cedar Falls commenced—the largest ground operation for American forces during the war. The project's aim was to eradicate the occupants of the tunnels of the nearby Iron Triangle, a Vietnamese Communist stronghold. The operation was at first deemed successful but, in the end, it was only a temporary setback for the Viet Cong. In 1969, American B-52s, which had previously been patrolling North Vietnam, flew southward and carpet-bombed the Cu Chi tunnels, exposing some of them to the surface. This event was more successful in slowing the Viet Cong than previous attempts.

The tunnels twisted and turned, snaking through the ground in South Vietnam, some of them even booby-trapped—with trapdoors that would send a soldier plummeting to the bottom of a pit filled with life-ending spikes.

Today the tunnels are still occupied by people. Except now, there are travelers instead of troops, sightseers instead of soldiers. About seventy-five miles of the tunnels have been preserved as part of a government-run war memorial site. They are now lit with electricity and, likely, visitors don't encounter the spiders and rats that once lurked in the underground passageways. Some tunnels have even been broadened to accommodate the height of Westerners and twenty-first-century humans. The underground conference room where war strategies, such as the Tet Offensive, were planned has been restored. And there's even a restaurant where tourists can eat what the ordinary Viet Cong subsisted on during the war.

When you arrive, tunnel guides, dressed in military fatigues or in black Viet Cong outfits, first show a short film of gruesome footage of the Vietnam War, called the "American War" in Vietnam: B-52s dropping bombs, snipers shooting away, grenades blowing up, and villagers running for their safety from American troops. And if you want to add to your Cu Chi experience, there is a gun range nearby that allows visitors to shoot AK-47s, just like those used in the war. Or you can take home a souvenir in the form of a Viet Cong military outfit or a toy gun.

Even with the commercialization of the tunnels, visitors walk away with a greater understanding of the Viet Cong experience during the American offensive. The tunnels are now recognized as a significant part of history.

A stove inside the living quarters of the Viet Cong's Cu Chi tunnels.

The *Dome of Light*, designed by Italian artist Narcissus Quagliata, incorporates symbols of Taiwanese cultural and political significance.

FORMOSA
Boulevard Station

KAOHSIUNG, TAIWAN

IS IT A MODERN CHURCH, AN ART INSTALLATION, A SUBWAY STOP? OR ALL of the above? The Formosa Boulevard Station, the busiest station in the Taiwanese city's metro network and an intersection of the city of Kaohsiung's Red and Orange Lines, is one of the world's most dazzling metro stations. The exterior of the station was designed by Japanese architect Shin Takamatsu, whose futuristic-looking structures include the Bidzina Ivanishvili Business Center in Tbilisi, the National Theater in Okinawa, and the Doshisha International Institute in Kyoto. Some say the four entrance/exit points to the subway station and their curvy glass design evoke two hands steeped together in prayer. Perhaps the architect used the theme of prayer because the Formosa Boulevard Station, the city's busiest subway stop, is where people congregate, where they come together.

Extending from subway wall to subway wall, the dome is one hundred feet across, consists of forty-five hundred glass panels, and covers an area of twenty-three thousand square feet.

But the impressive interior is only the beginning. Opened in 2008, Formosa Boulevard Station is not just a place where commuters change subway lines. Step inside, and you'll see why it is often ranked as the most beautiful subway station in the world. It's all about the Instagram-friendly *Dome of Light*, the largest glasswork on the planet. The dome was designed by Italian artist Narcissus Quagliata, who is famous for stained glass windows in San Francisco's Grace Cathedral, the glass dome in Rome's Santa Maria degli Angeli e dei Martiri, and the largest single-image stained glass window in the world, located in the Church of the Resurrection in Leawood, Kansas. It took four and a half years to construct, each panel designed and created at Quagliata's studio, manufactured in Germany, and then put together at the subway station. Extending from subway wall to subway wall, it is one hundred feet across, consists of forty-five hundred glass panels, and covers an area of twenty-three thousand square feet. The window comprises three different types of glass from the Venetian island of Murano.

The images on the back-lit glass dome tell the story of the cycle of life in four separate parts, each with its own title: "Water: the Womb of Life," "Earth: Prosperity and Growth," "Light: The Creative Spirit," and "Fire: Destruction and Rebirth."

The *Dome of Light* was a sensation from day one. The city began giving tours of the station six months before it officially opened because of growing public interest in what the Italian artist had created. Originally, the tours were limited to a small number of people and were planned for

The Formosa Boulevard Station graces passers-by not only with the changing colors of the *Dome of Light* during their commute but also with the installation's symphonic music.

only one weekend in March. But demand intensified, and they extended it to nearly every weekend until the station officially opened in September 2008.

Standing under the *Dome of Light*, with its colorful stained glass and images of birth and destruction, you can see why the public was eager to see it. After all, you'd be forgiven if you mistakenly thought you'd just stepped into a cathedral. The images on the back-lit glass dome tell the story of the cycle of life in four separate parts, each with its own title: "Water: the Womb of Life," "Earth: Prosperity and Growth," "Light: The Creative Spirit," and "Fire: Destruction and Rebirth."

The various hues of the rainbow that make up the installation glow and shine and pulse. Two columns located in the center of the large subway space, one red and one blue, constantly change colors. And symphonic music blasted from speakers enhances the dramatic effect.

The glasswork represents life and death but also Taiwan's budding democracy. The station is named for the Formosa Incident, otherwise known as the Kaohsiung Incident or the Meilidao Incident. On December 10, 1979, *Formosa Magazine* organized a demonstration for Human Rights Day. It seemed innocuous enough. But it was really a demonstration to raise awareness for democracy in Taiwan and eventually resulted in Taiwan's prodemocracy movement.

SUBWAYS

NEW YORK CITY, USA

LONDON, England

BEIJING, China

MOSCOW, Russia

PRAGUE, Czech Republic

THE SUBWAY. THE METRO. The Underground. Whatever you call it, these urban mass transit trains are a wonder of the twentieth century, and a city is not world-class without one. These subterranean trains are more than just people movers. In some ways, they are the soul of a city. What would New York or London be without their expansive subway and underground systems, respectively? Subways help create vibrant, bustling metropolises in ways that aboveground transportation—from taxis to buses to Uber—never will.

In 1904, the very first New York City subway ride took place, leaving from the City Hall station. Commuters may not have spent much time looking around, but those who did were in store for a visual treat: chandeliers hung from the vaulted ceiling of the station, where colorful tile designs, created by Rafael Guastavino, dazzled the eye. There were sculptures by Gutzon Borglum of Mount Rushmore fame. The architects, Christopher Grant LaFarge and George Lewis Heins, also built the Cathedral of St. John the Divine. Very few people now have uttered the words "New York subway system" and "beautiful" in the same sentence, but if more stations resembled the old City Hall station, then Gotham's subway system would certainly garner more compliments. Sadly, City Hall station was decommissioned because the nearby Brooklyn Bridge station had

Built in 1904, the City Hall subway station was the first to be opened in New York City.

access to both local and express trains (while City Hall station was just local). And so on December 31, 1945, the station was closed. Fortunately, it was not demolished, and the city gives occasional guided tours of City Hall station throughout the year. But the New York City subway marches on. And despite it not being particularly aesthetically pleasing, with its overheated platforms in midsummer and overcrowded trains during rush hour, it's still incredible that anyone can traverse the bustling Big Apple underground for just $2.75. The longest ride on the New York City subway system is the A train: going from Far Rockaway in Queens to 207th Street in northern Manhattan, a rider travels thirty-one miles. The deepest part of the subway is the 191st Street station at 180 feet below the street. And if you laid all the track together, end on end, the rails would stretch from New York City to Chicago.

The very first subway—or metro or underground—was in London. Opening in 1863, it was called "the great engineering triumph of the day" by the *Times of London*. The first train chugged out of Paddington Station (then called Bishop's Road) on its way to Farringdon Street on January 10, 1863. On that day, thirty thousand Londoners tried it out. In its long history, the Tube, as it's referred to in the local parlance, has seen a lot of change. Early trains, for example, were powered by locomotive steam engines. Then in 1890, the first electric trains were used. Eighteen years later, electronic ticketing machines were installed. And in 1911, London's first escalator began shuttling people up and down from the platform at Earl's Court station. The year 1929 saw the end of the last manually operated doors, which were replaced with automatic doors. And then in 2007, the Underground carried one billion passengers in a year's time. Today, almost five million people per day still "mind the gap" on the Underground's eleven lines, which stretch 250 miles in total.

WHILE LONDON IS HOME to the very first underground metro system, say *ni hao* to the busiest in the world, the Beijing Subway, which

The Baker Street Tube station in London, England.

An entrance to the Beijing Subway, which has one of the largest annual riderships in the world.

has an annual ridership of over three and a half billion people. Nine million riders per day take the subway in the Chinese capital, equal to the entire population of Switzerland! It boasts 357 miles of track, making it only second to Shanghai's as the longest subway system in the world. Opened in just 1969, the metro system was the first in mainland China. The initial line had perpetual

Beijing Subway passengers in transit.

electric problems, and a second line would not be built until ten years later. So the metro system was made up of only two lines until it was expanded in 2002. And oh, did it expand. In fact, it's still growing. Soon, the city hopes to have 621 miles of subway track down and projects there will be 18.5 million trips per day. This is all potentially good news for a city with one of the worst air-pollution problems on the planet. It's also good news for those who need to traverse the city while avoiding the monstrous traffic. Beijing is a fast-growing metropolis, spread out like Los Angeles. Of the nineteen lines, two of them are circle lines, subway routes that ring outer parts of the city.

AND NOW TO POSSIBLY the world's most beautiful subway system. Moscow's metro system was christened in 1935 when Joseph Stalin became the first rider on the subway. According to *Pravda*, in the days leading up to the opening, the train driver took practice runs with a dummy of Stalin seated in one of the carriages. The metro's 206 stations and 210 miles of track belie the real eye-popping details of Moscow's underground transport system: many of its stations are bedecked with chandeliers, neobaroque moldings, high vaulted ceilings, marble arches, stained glass windows, and rows of colonnades. In fact, forty-four of the stations are

The Hall of Komsomolskaya
subway in Moscow.

designated as Russian cultural heritage sites. As they should be. Take Kiyevskaya station, for example. The long, narrow platform is a riot of color and movement: the curvy moldings, the intricate frescoes and mosaics that depict life in Ukraine, and the marble walls look like they could have been snatched from a baroque opera house. Even the newer stations are a feast for the eyes: Dostoyevskaya station, built in 2010 and named for legendary Russian author Fyodor Dostoyevsky, is beautiful in its stark minimalism. Murals sprinkled throughout the station show scenes from his novels *The Idiot* and *Crime and Punishment*. This station is a must for any fans of Russian literature. And for Moscow Metro fans, there's good news: the city is currently expanding the system. Soon they will have added sixty-two more stations. One can hope that at least some of the new stations will live up to the beauty of those already in existence.

MOVING OVER TO Central Europe, Prague's subway system can't compare to the above systems in their superlative lengths and sizes—there are only three lines with sixty-one stations and just forty miles of tracks—but this smooth and clean metro deserves a mention because of the depth of its stations. Built by the Soviets in 1974, some of the stops were meant

Náměstí Míru station in Prague, Czechoslovakia.

to double as bomb shelters in case the Cold War became a nuclear war. The escalator at the Náměstí Míru station, for example, is the longest in the European Union, at 285 feet and 533 steps (the deepest metro station in the world, by the way, is Kiev's Arsenalna station at 345 feet); the trip from top to bottom takes two minutes and fifteen seconds. During the Communist era, which ended in 1989, many of the stations had socialist-referencing names, including Moskevská (Moscow) station, which is now Andel. While the city has removed many of the Communist-era relief sculptures around this Czech city, the Andel subway station still boasts some plus-size images of heroic and triumphal proletariat looking with pride and optimism to the future.

The subway stations in Prague,
Czechoslovakia, were built deep
enough to also serve as bomb shelters
in case of a Cold War nuclear attack.

From the
BIG APPLE
to the
ANDEAN
HIGHLANDS

Some of the flora and fauna planned for the Lowline.

THE LOWLINE

NEW YORK CITY, USA

NEW YORK CITY HAS AN ABUNDANCE OF WAYS TO EXPLORE THE literal underground city: subterranean bars and restaurants, cheese caves in Brooklyn, crypts below Saint Patrick's Cathedral, tunnels that run underneath McCarren Park in Williamsburg, and, of course, the extensive subway system.

Now we can add a park to the list. Meet the Lowline, an ambitious project on Manhattan's Lower East Side that's transforming the long-abandoned former Williamsburg Bridge Trolley Terminal into the world's first underground park.

As most people would guess, turning a dank, dark trolley terminal into a vibrant, well-lit space with growing vegetation isn't easy. For this reason, the Lowline is an innovative technology- and design-driven project, named by *Time* magazine as one of the top twenty-five inventions of 2015.

Dan Barasch, cofounder of the Lowline, said, "It's not that often that you can transform a forgotten piece of real estate in a city like New York and turn it into a magical public space."

This "magical space" was originally created in 1908 as a terminal for passengers wanting to take the trolley across the Williamsburg Bridge. It shut down in 1948 and was abandoned until 2008.

A few years ago Lowline cofounder James Ramsey, an architectural designer, got access to the space. His mind started to race: How do you bring natural light and flourishing plant life to a cavernous, buried room that has no windows or skylights?

The Lowline Lab was composed of more then three thousand plants and dozens of unique varieties, spread across one thousand square feet.

That's when Ramsey and Barasch got to work. Their team created a series of remote skylights with solar panels that track the sun all day and direct the light underground through a system of fiber-optic cables and mirrors, concentrating it to thirty times the brightness of the sun. A thirty-foot-wide anodized aluminum solar canopy disseminates the light throughout the one-acre area. The goal is to grow more than thirty thousand plants of sixty different species.

And how do they know this will work? They've already tested it in the Lowline Lab, an aboveground site in an abandoned market where they simulated an underground space two blocks from the actual Lowline. There they did a series of controlled experiments with solar panels and skylights and found success in growing thriving plants, including herbs and fruits.

If it all sounds like an urban science-fiction tale about future humans living underground, you may be right. E. M. Forster's short story "The Machine Stops," about a future society that lives completely under the earth, comes to mind. But if you put the Lowline into the context of New York City, it makes sense. The city, as one of the most compact urban environments on the planet, has always needed to get creative when it comes to living and leisure space. Now instead of skyscrapers reaching up, innovators are looking down.

The Lowline, at least in its name, is a nod to the High Line, another creative use of an abandoned site in the Big Apple. An erstwhile elevated railway that ran down the west side of Manhattan, the elevated park became an instant success when it opened in 2009, as crowds flocked to the inspired landscape design, food and drink kiosks, and public art. The Lowline is the underground version of this, and the cofounders hope it will be an instant classic, just as the High Line has become.

But unlike other parks in New York City, including the big daddy of all urban parks, Central Park, the Lowline is not your typical swath of green grass where families and friends barbecue on weekend afternoons and throw the ball around. Instead, the Lowline is something closer to a botanical garden. In addition to grass, the Lowline's vegetation could include ferns and mosses, as well as edible flora, such as tomatoes, pineapples, and strawberries, all thanks to the ventilation system and the beaming in of natural light. A smaller-scale model of the park reveals a simulated canyon made of plywood, a replica of the famed Antelope Canyon in Arizona.

One of the most compact urban environments on the planet, New York has always needed to get creative when it comes to living and leisure space. Now instead of skyscrapers reaching up, innovators are looking down.

Is the Lowline, with its cutting-edge lighting that allows plant life to flourish, a test run for further life underground? Or preparation for when we abandon the increasingly unlivable Earth for another planet? Until then, it will just be a great way for New Yorkers to get a little more nature in their lives.

A solar canopy, designed and constructed by engineer Ed Jacobs, spreads the collected sunlight across the space in order to sustain the plant life.

Vistors to Louisville, Kentucky's Mega Caverns partake in one of the many activities that can be found inside the underground complex.

MEGA CAVERN

LOUISVILLE, KENTUCKY, USA

I F YOU WOKE UP RECENTLY AND THOUGHT, *I REALLY WANT TO ZIP-LINE in a dark, underground space,* then you're in luck. Located under the Louisville Zoo and the Watterson Expressway, Mega Cavern boasts the only underground zip line in the world. Sure, one and a half million people gravitate to this Kentucky town every year for its famous horse race. But underground zip line aficionados (and the people who love these thrill seekers) race straight for Mega Cavern. The Mega Zip is a one-hundred-foot drop on a zip line straight into the blackness, making this attraction very unique. And very adrenaline inducing.

The one-hundred-acre limestone space is the largest structure in the state of Kentucky. Ralph Rogers opened the cavern in the 1930s to mine the massive limestone deposits. The famed Cincinnati Arch is made from stone that was excavated in the cavern. Mining stopped in the 1970s, and the cavernous space sat abandoned for a couple of decades until a private owner with a grand vision for the cavern took over in the late 1980s. A ten-ton heater dehumidifies the fifty thousand square feet of underground space. Opened to the public in 2009, the cavern's seventeen miles of halls and corridors one hundred feet below street level mean it functions as more than just a zip line into the abyss. Mega Cavern even houses a business park, and about a dozen companies have offices here. It also offers tours: visitors can get a guided excursion in a topless tram carriage pulled by a Jeep Wrangler. Around the holidays, guests for the light show can drive their own cars through the cavern to gawk at the world's largest display of holiday lights, where over two million twinkling bulbs, making up 850 different characters and forms, dazzle the eyes. But if pedaling is more your style, the cavern is also host to the world's largest mountain biking track, with twelve miles of trails. Hundreds of pounds of sand and dirt were brought into the Mega Cavern to help form ramps and jumps and forty-five different trails for various types of bicycle riding, including BMX and cross-country. And if you just can't muster up the energy to pedal around on a bicycle, you can also take an electric bike tour of the cavern, a ninety-minute ride that takes you to some usually off-limits

With excursions like zip-lining, mountain biking, and an aerial ropes challenge, the Mega Cavern provides no shortage of adventure for thrill-seekers.

parts of the structure. There's also an aerial ropes challenge, where the adventurous can climb walls made of rope, cross rope bridges, and swing like a monkey.

According to CNN, the cave has a few secrets too: its cool, dry atmosphere makes it perfect for storing things, such as secret government documents, which apparently are hidden somewhere in the four-million-square-foot space. In fact, Mega Cavern is one of eight facilities in the United States that has been authorized by the national archives and records administration to archive federal documents. After the terrorist attacks of September 11, 2001, the US government felt it needed to store its records and files in the safest possible place. They decided that place was the Mega Cavern. Who knows what secrets it may hold? It also had a brief flirtation with Wall Street, as it was set to house a backup supercomputer linked to the financial markets, with fiber-optic cables stretching 750 miles from Louisville to southern Manhattan, but plans were shifted at the last minute. Other typical objects that are stored underground here include classic cars—some of which have rarely seen the light of day—and boats, the main advantage being that owners don't have to weatherize them since the climate inside the cavern is always the same.

After the terrorist attacks of September 11, 2001, the US government felt it needed to store its records and files in the safest possible place. They decided that place was the Mega Cavern. Who knows what secrets it may hold?

The average fifty-eight-degree temperature makes it ideal for storing things like chocolate and bourbon, which are hidden in the cavern. And Hollywood studios store their old films here, the fate of these classics resting on the cool climates of the Mega Cavern. In addition to all that, there are also worm farms lurking within the oversize grotto.

During the Cuban Missile Crisis in October 1962, this space was going to act as storage for something else: humans. At least fifty thousand people can fit inside, and the Mega Cavern was set to become the biggest fallout shelter in the United States. Fortunately, the chocolate and bourbon remain, and the only humans in Mega Cavern today are here for recreation and curiosity. Before visitors take the tram tour through the massive cave, they are shown Cold War–era government propaganda films just to get them in the right mood.

And after your free fall into darkness via the zip line, you can indulge in that other Kentucky favorite: bourbon. You may need a shot of it after spending some time in the Mega Cavern.

Light shines through opaque tiles that make up parts of the Seattle Underground's ceiling.

SEATTLE Underground

SEATTLE, WASHINGTON, USA

O N JUNE 6, 1889, JOHN EDWARD BACK, A YOUNG APPRENTICE cabinetmaker, was at work at the Clairmont and Company Cabinet Shop on the corner of Front and Pearl Streets (now First Avenue and Madison Street). He was boiling glue and at one point became distracted. At 2:39 p.m., he made an abrupt move and accidentally knocked over the pot of hot glue. The glue, which was mixed with grease, spilled on the wooden floor and ignited a small flame. Mr. Back's first reaction was to douse it with water. Bad mistake. This only made the grease fire spread. Soon the building was up in flames. What made matters worse was that the fire department chief was on vacation. And while there were certainly enough members of the volunteer fire brigade who rushed to the scene, the amateur fire fighters used too many water hoses at once, thus lowering the water pressure, which barely pushed the water out of each hose. The flaccid spray of water was not able to reach much of the fire. Even worse, the wind was strong that day, feeding the flames. And with that, the fire spread to the adjoining building and then the next and the next. The buildings were all made of wood and went up like tinder. In the end, over two dozen blocks were reduced to complete rubble and ash. It was a perfect storm of circumstances and devastated the Queen City. The following day the *Los Angeles Daily Herald* ran a front-page article about the fire with the headline, "Seattle in Ashes." Seattle's main newspaper, the *Seattle Post-Intelligencer*, couldn't report on the story because their presses had burned to the ground.

The fire was a significant moment in the history of this city in the Pacific Northwest. In the rebuilding effort, city planners made several key decisions: all new buildings would be made of stone or brick. They would also rebuild the city one or two stories higher than the original: prefire Seattle was constructed on soggy tidal flats and was prone to muddy flooding whenever it rained a lot (which was often). An added benefit: by raising the city ten or twenty feet, gravity-assisted toilets would flush easier.

So the city began to re-create itself. They built walls and pillars to support the new streets, thus creating underground alleyways that would sit buried for decades. In some parts of the city, they raised the streets by twelve feet and in other parts as much as thirty feet. Even before the new streets and sidewalks were complete, though, new buildings sprang up. Building designers knew the first floor would eventually become the basement, so the lowest floor of these new buildings was unremarkable while the next floor up—the new ground floor—was flush with design elements. While the new city streets were being constructed, the now-subterranean streets were still used, lit by glass skylights installed on the newly raised surface of Seattle. One other crucial problem was that the city ran out of money after a few years and couldn't complete the newly raised areas. Or at least not the sidewalks. So, in some cases, if you wanted to cross the street, you had to go down a ladder into the old street and then back up a ladder to complete your journey to the other side. This

After being closed up in 1907 to prevent the spread of disease, the Seattle Underground reopened to visitors in the 1960s.

was hazardous to Seattleites, as many would fall into the pits where sidewalks should have been. At one point the city's morgue had seventeen people who had died from "involuntary suicide," having accidentally plummeted into the deep gaps.

About ten years after the fire, a wave of newcomers came through Seattle. As people heard about the Yukon gold rush, wealth-seeking individuals gravitated north, most of whom stopped in the Queen City. And, as happens with anything related to a gold rush, debauchery followed. In Seattle, this debauchery often took place underground. In this case, literally *and* figuratively. In the old, now-buried streets of Seattle, speakeasies, brothels, opium dens, and casinos popped up.

And then in 1907, the city closed up the underground portion of the city. There was fear of an outbreak of bubonic plague, which would have devastated a city on the verge of hosting the 1909 world's fair.

Seattle Underground has its own place in pop culture, making cameos in the 1972 made-for-TV movie The Night Stalker *as well as a 1976 episode of* Scooby-Doo.

In general, Seattle's former self, its buried past, was largely forgotten about. And then in 1964, Bill Speidel, a columnist for the *Seattle Times,* received and published a letter from a reader asking what he knew about what was underneath then-derelict Pioneer Square. Speidel didn't know much but promised he would do some research on it. After digging around in the archives, he responded to the reader via his column, asking to meet in Pioneer Square and saying they would go on tour of the space beneath the square. The reader turned up. So did three hundred other people. After that, Speidel realized that tourists might be interested in underground Seattle, and he began doing tours, charging tour goers just one dollar for the experience. He got permission from building owners to allow him to escort the curious and adventurous below the streets of the city, where he would tell them extravagant tales of Seattle's past. Some years later, Mr. Speidel started an R-rated version of the tour, going into more lurid detail about the brothels and gambling halls that sprang up after the new city was built. Speidel passed away in 1988, but his tours still exist. The guided walks take visitors through the few blocks of the old city that are still preserved, giving the history of prefire Seattle and showing relics from that era, including old signage and sidewalks.

Seattle Underground has its own place in pop culture, making cameos in the 1972 made-for-TV movie *The Night Stalker* as well as *Scooby-Doo* in the 1976 episode "A Frightened Hound Meets Demons Underground." (Those meddling kids!)

And speaking of all things demonic and ghoulish, there is—naturally—reported paranormal activity in the underground city. An old bank vault and a spot where prostitutes lingered are two of the most popular hangouts for apparent ghosts.

The entrance to the Cheyenne Mountain Complex, a military bunker where workers perform their duties thousands of feet below the ground.

CHEYENNE MOUNTAIN COMPLEX

CHEYENNE MOUNTAIN Complex

COLORADO SPRINGS, COLORADO, USA

HUMANS WOKE UP TO A DIFFERENT WORLD ON AUGUST 29, 1949. That's because the day before, the Soviets detonated a nuclear bomb in Kazakhstan, thus ramping up the Cold War and accelerating the arms race between the United States and Soviet Union.

In response, the United States and Canada formed North American Aerospace Defense Command, or NORAD, a command center that would be connected to a vast, earth-spanning early-warning network of radars and satellites, which tracks activity, mostly of the aerial variety, in order to protect the United States and Canada from nuclear threats. For the first few years of NORAD's existence, its headquarters was a shabby 1920s hospital building in Colorado Springs, Colorado. Not exactly the most well-protected site if you're concerned about being wiped out of existence by your global arch enemy, the Soviets. The command center needed a new HQ, something that could withstand, say, a nuclear bomb. Location possibilities included an old mine shaft and an underground bunker. When NORAD officials looked around Colorado Springs, they saw only towering mountains. Then someone had a light bulb moment: What if we put it in a granite mountain in the Rockies?

The result was Cheyenne Mountain Complex, named after the Native Americans of the area. The ninety-five-hundred-foot-tall mountain, sometimes called "America's fortress," houses fifteen two- and three-story buildings, all connected by corridors. The entire space inside the mountain is the size of five football fields. Not that you'll ever see the inside with your own eyes: security here is tighter than at the Pentagon.

But if you were one of the few hundred people who work here—a mixture of American and Canadian military personnel and ordinary civilians—you'd enter the complex by taking a drive in a fourteen-hundred-foot tunnel before coming to one of the two blast doors. The three-foot-thick hydraulic-run bank-vault-like door is made of twenty-five tons of steel. And given such a heavy beast of a door, it's incredible that it can be closed in half a minute. But the door usually remains open. If it's closed, then you know the country is under attack or there's a serious national security alert. The last time they closed the door? September 11, 2001.

Work on the facility began in the late 1950s, and by 1959, they'd hit a roadblock: the entrance to the mountain would be one thousand feet up the mountainside, and there was no road to access it. So the first order of business was to build a four-mile road that went up the mountain. After that, they began digging into the granite in 1961. Three crews working eight-hour shifts labored around the clock six days per week, blasting into the mountain. They used 1.5 million pounds of dynamite, excavating 693,000 tons of granite from the mountain to create the caverns that would become this important Cold War command center. At the same time, in August 1961, the Berlin Wall was erected, and in October of that year the Soviets detonated the Tsar Bomba, a fifty-seven-megaton hydrogen bomb that is to this day the most powerful human-made explosive in history (it was four thousand times stronger than the atomic bomb dropped on Hiroshima). To Americans, the NORAD command center was more necessary than ever. In 1967, six years and $142 million later, it was complete. (Adjusted for inflation, the complex would cost $18 billion to build today.) The United States (and Canada) would now have a supersecret base inside a mountain—likely the most secure place on the planet—which seemed like something straight out of science-fiction novels and films.

The ninety-five-hundred-foot-tall mountain, sometimes called "America's fortress," houses fifteen two- and three-story buildings, all connected by corridors. The entire space inside the mountain is the size of five football fields.

And so what exactly went on inside this mysterious mountain? In the command center, round-the-clock workers monitored the Soviet Union, particularly its missile and military activity. But after the Cold War ended, personnel inside the structure took on new roles, like monitoring the twenty-three thousand objects and satellites that orbit the earth, checking to see if there are any outer space threats, human made or natural. They also watch for missile activity from rogue countries, and after the September 11 terrorist attacks, they began monitoring every commercial flight in the United States. If a plane appears to be hijacked, they can get fighter jets in the

A serviceman passes through a twenty-five-ton door that secures the complex.

air within seconds to find the plane. And given Cheyenne's top-secret activity, who knows what else is going on in there?

The Cheyenne Mountain Complex is really a city in a mountain: there's a restaurant—called, fittingly enough, the Granite Inn—serving up burgers and sandwiches, as well as a gym, a basketball court, and a mini-mart selling snacks. There are four reservoirs holding six million gallons of water for cooking and drinking and a fifth reservoir containing 510,000 gallons of diesel fuel. At one time, there were up to two thousand people who worked here; today, only about two hundred employees clock in.

Everything inside the mountain was designed to withstand a nuclear blast: the pipes running through the interior are made with a flexible material. Underneath the floor are 1,319 springs—three feet tall, two feet wide, and weighing a ton each—whose function is to absorb any kind of blast from the outside, or even an earthquake. The springs allow the entire complex to sway twelve inches in either direction. At its peak, the mountain was completely self-sufficient: in case of a nuclear attack, the complex was designed to hold hundreds of people inside for a month.

> *Underneath the floor are 1,319 springs—three feet tall, two feet wide and weighing a ton each—whose function is to absorb any kind of blast from the outside, or even an earthquake. The springs allow the entire complex to sway twelve inches in either direction.*

Then in 2006, the government minimized operations at the complex. NORAD is now headquartered at nearby Peterson Air Force Base, and today the Cheyenne Mountain Complex houses a training and alternative command center for NORAD. But the security and command bunker still lives on in our imaginations, thanks in part to its appearance in pop culture. The command center was a setting for the 1983 movie *WarGames,* and the entrance to the mountain was featured in *Independence Day.* In Stanley Kubrick's 1964 film *Dr. Strangelove: or How I Learned to Stop Worrying and Love the Bomb,* the command center is very similar to Cheyenne's. In the TV show *Stargate SG-1*, the stargate command is located inside the mountain for the entirety of the series. Today, there is a broom closet inside Cheyenne with a nameplate that reads STARGATE COMMAND.

PEACH SPRINGS, ARIZONA, USA

THE VAST CAVERNS UNDERNEATH THE ARIZONA DESERT, CALLED THE Grand Canyon Caverns, were discovered by a cowboy and woodcutter named Walter Peck as he walked to a poker game in 1927. After he'd almost fallen into a funnel-shaped gap in the desert floor, he wondered what might lurk below. So the next day he was back with friends, a lantern, a rope, and a lot of curiosity. He lowered himself down to discover a vast network of limestone caves that snakes its way through the netherworld landscape. At first Peck thought he'd hit the jackpot—a gold mine—so he bought the property and began exploring, hoping his fortunes would soon change. Unfortunately for him, the sixty-five-million-year-old limestone grotto contained no gold.

But after investing in the land, Peck was intent on reaping reward from the caverns anyway. Lucky for him, Grand Canyon Caverns is unusual in that it's a dry cave—only 3 percent of caves in the world are dry. With no bacteria or other organisms that can break down dead bodies, and with the temperature being a perpetual fifty-seven degrees—and only 2 percent humidity—conditions are ripe to encourage preservation and mummification.

And so, Peck began promoting the caverns by promising the sight of mummified cavemen and charged brave visitors twenty-five cents to be lowered down on a rope, a lantern in hand. The "cavemen" in question were two Hualapai Indians who fell into the caverns in 1918. The bodies were eventually returned to their native burial

A view of the Grand Canyon, concealing its many underground spaces.

grounds, but there were other artifacts in the caves that kept early twentieth-century tourists intrigued, and in 1936, Peck had a more conventional entrance, by way of a metal staircase, installed.

When Peck first began exploring the cavern, he found the remains of various animals that had fallen into it, become trapped, and died, their bodies mummifying over the centuries. Take a tour of the caves today, and you can see some of these animals still on display. There's "Bob," a bobcat that experts believe fell into the cave around 1850. With a broken hip from the fall, Bob had also inhaled vast amounts of limestone dust, which filled his lungs until he suffocated. Then there's a fifteen-foot sloth who fell into the cavern about twelve thousand years ago. The sloth tried unsuccessfully to climb out; if you look at the ceiling just above where the sloth stands today, you can see her claw marks, signs of her desperate attempt to escape.

The caverns formed millions of years ago. When what is now Arizona was covered by an ocean 350 million years in the past, the skeletons of sea animals created calcium deposits, which eventually turned to limestone. About thirty-five million years ago, water erosion helped form the caverns. One of the more amazing facts about the Grand Canyon Cavern is that much of it is still unexplored. Geologists didn't actually know how long the caverns were until they did a test, shooting smoke into the caverns to see if it would pop up somewhere as it traveled through the cave. It did: three weeks and sixty miles later, in the Grand Canyon.

Twenty-first-century visitors can now descend into the cavern via a modern elevator, first installed in 1962, which took three years to construct: two years to dynamite the shaft and a year to install the lift. It takes about fifty seconds to drop the twenty-one stories to the bottom of the cavern. The first room is a massive space, the size of two football fields. On the seventy-five-minute tour, guides regale guests with the history of the cavern, which includes some Cold War–era artifacts: like many cavernous spaces in the US, this one was going to double as a bomb shelter during the Cuban Missile Crisis in 1962. Authorities stocked the cavern with rations. And because of the preservation-friendly climate conditions of the cave, the rations are said to still be okay to eat.

Before Peck's discovery of the site, at least eight people had perished in the caves, most of whom unexpectedly plummeted through a hole or crevasse on the surface of the desert and were not rescued or could not escape. And so, naturally, there is a paranormal aspect to the Grand Canyon Caverns, as ghost hunters and other astute individuals have claimed to hear whispering voices, some attributing them to the ghosts of the Native Americans who were once here. Others have supposedly seen transparent figures dancing in the distance. There has been enough activity that even the TV show *Ghost Adventures* dedicated an episode to the possible supernatural activity in the caverns.

Visitors to the Grand Canyon can stay twenty-two floors belowground in the Cavern Suite.

And if you like the caverns so much (and aren't too spooked by a possible ghost sighting) that you want to spend more time two hundred feet below the earth, you can sleep there in the Cavern Suite. You just need $850 (per night) and an inclination to learn the true meaning of the term "deep sleep." Billed as the deepest, darkest, and quietest hotel room in the world, the Canyon Suite at the Grand Canyon Caverns is 220 feet below the earth and features everything from a thirty-two-inch flat-screen TV to a coffeemaker to a record player. Sixty-five-million-year-old limestone walls enclose the plus-size room, which is two hundred feet in diameter, with a seventy-foot-high ceiling. After checking in, you'll be escorted to the room, shown the amenities, and then left alone until the morning, when breakfast arrives. That is, if you check in after 4:00 p.m., when the last tour finishes, and check out before 10:00 a.m., when the first tour of the day begins. You're welcome to stay longer, but you'll just have to be okay with being gawked at by strangers: from the late morning to the late afternoon, the tour of the caverns shows off the suite to visitors.

For those who can't bear the idea of spending the night in the cavern, consider dropping in for a meal. In August 2017, the Cavern Grotto restaurant fired up its burners for the first time. Guests get ninety minutes for lunch and two hours for dinner to enjoy this culinary cave experience.

Pre-Columbian pyramid
engravings at Teotihuacán.

TEOTIHUACAN

MEXICO CITY, MEXICO

AFTER A PARTICULARLY HEAVY RAINSTORM IN MEXICO CITY ONE day in 2003, archaeologist Sergio Gómez decided to go to the city's pyramids at Teotihuacán, located thirty miles from the bustling Mexican megalopolis, hoping there wasn't too much damage. Souvenir stands were knocked over. Swaths of land were lightly flooded. And, at the base of one of pyramids called the Temple of the Plumed Serpent, a three-foot-wide sinkhole had formed. The sinkhole revealed something incredible: a hitherto unknown passageway.

Gómez, who knew the pyramids better than anyone, remembered the day in the 1980s when they discovered 137 bodies under the pyramid (probably sacrificial victims from when it was built). But as far as he was concerned, there was nothing else below it. When he got down into the sinkhole, though, he saw a corridor made by human hands and, at both ends of the hallway, two massive stones blocking each entrance/exit. Were archaeologists about to get a new perspective on Teotihuacán, the ancient and storied pre-Columbian city?

Teotihuacán, also called the "Birthplace of the Gods," consisted of twenty-five square miles of temples, markets, palaces, apartment buildings, plazas, and avenues. It had springs and canals that carried water through the city from the nearby San Juan River. And two millennia ago the population was between one hundred thousand and two hundred thousand people, making it one of the most populous cities on the planet (and certainly the largest in the pre-Columbian

Americas). At the center of this vast urban settlement were the three pyramids, the Temple of the Sun, the Temple of the Moon, and the Temple of the Plumed Serpent, all lining the sacred road, the Avenue of the Dead.

The Temple of the Plumed Serpent, referred to as the Temple of Quetzalcoatl by the ruling Aztecs who lived in the area from the fourteenth to the sixteenth centuries, is bedecked with iconic relief sculptures of the heads of feathered snakes, each one

Workers unearth a tunnel beneath Teotihuacán.

with its mouth open, bearing teeth. Some experts have theorized links between the serpents and the creation of time and the cosmos.

As it turns out, the tunnel that Gómez found, 330 feet right under the Temple of the Plumed Serpent, had glittered, metallic dust affixed to the ceiling, leading some to believe the world under the tunnel represented a passageway through the cosmos and space. Scholars were aware of a similar tunnel and a chamber under the nearby pyramid, the Temple of the Sun, but it had long been empty, thanks to tomb raiders. Experts made a digital map of the corridor under the Temple of the Plumed Serpent, but they still were not able to budge the boulders on either side of the corridor, which they suspected were placed there two thousand years ago.

And so in 2009, Gómez and his colleagues began chipping away at the rock, hoping their work would lead to the discovery of a subterranean chamber that would hold more keys to understanding the Teotihuacános who inhabited the city from the second century BC to the seventh century AD.

The tunnel that Gomez found had glittered, metallic dust affixed to the ceiling, leading some to believe the subterranean world under the tunnel represented a passageway through the cosmos and space.

Five years later, in 2014, the team of archaeologists finally broke through to discover three secret chambers sixty feet below the base of the pyramid. Gómez was hoping to find the tomb of powerful Teotihuácan rulers. Alas, not so. There wasn't much in the rooms, save for a few statues, some jewelry, large conch shells, jade ornaments, and mercury, which the experts speculated was used to simulate sacred lakes and rivers.

For now, the corridor and chambers beneath the Temple of the Plumed Serpent are still off-limits to tourists. Someday visitors will be able to wander the netherworld beneath the temple, but until then, the team of archaeologists will continue their research and excavations.

Meanwhile, in 2017, archaeologists made another discovery: a tunnel underneath the Temple of the Moon, the second-largest pyramid in the area (after the Temple of the Sun). The tunnel, about thirty feet below the ground, mirrors the tunnels of the other two pyramids. Though nothing significant was found in either of the first two pyramids, Sergio Gómez hopes the third time will be a charm when he finishes excavating the tunnel underneath the Temple of the Moon.

The Tello Obelisk at Chavín de Huántar, a UNESCO World Heritage site in Peru.

ANDEAN HIGHLANDS, PERU

LOCATED 10,500 FEET ABOVE SEA LEVEL, AND 150 MILES NORTH OF Lima, Chavín de Huántar is a fascinating, relatively off-the-radar archaeological site that gives clues to perhaps South America's oldest civilization. And most of those clues are underground.

Chavín was first discovered in the late nineteenth century. But before anyone could really explore and study it, the three-thousand year-old complex was buried under massive mudslides. It wasn't until just after World War II when it was fully uncovered. And it wasn't until the 1990s, when archaeologists from Stanford University began spending time at the site, digging and measuring, that we began to know more about the site's history. The Stanford team, for example, figured out that Chavín was much older than anyone had previously believed, determining that its construction began around 900 BC, and that it was built in at least fifteen different stages over a few hundred years. Chavín de Huántar is two and a half millennia older than Machu Picchu, the ancient Inca citadel that sits hundreds of miles to the south and draws far more tourists than Chavín.

Not a lot is known about the Chavín people, but archaeologists do know this: the complex they left us has been instrumental in influencing civilizations that came after them. In particular, the temple's location is significant. It's set between the western and eastern ranges of the Andes Mountains at a spot with relatively easy passage between the Amazon jungle to the east and the desert coast to the west, allowing pilgrims to come from far and wide to pay their respects here. It's also near the joining of the Huachesca and Mosna Rivers, a natural confluence that ancient humans considered spiritually significant.

At first sight Chavín doesn't look like much: there's a verdant central square, large mounds of earth, and the remnants of various buildings, including part of a staircase that once led to a forty-foot-tall structure. There were also two chief buildings made of granite: a U-shaped Old Temple, probably built near 900 BC, and the New Temple, constructed around 500 BC.

But go underneath, to the subterrestrial chambers and passageways, to get a real clue about the people who built this intriguing stone complex. Some of the passageways have evocative names: Gallery of the Offerings, Gallery of the Bats, Gallery of the Madman, for example.

Archaeologists are still not one hundred percent certain what all of the underground rooms were for, and digging continues to this day. But they have found some intriguing items and have determined that one major function of the subterranean chambers was to serve as a place of worship. One item that existed below the temples is El Lanzón, a fifteen-foot white granite obelisk that is intricately carved with an

Chavín de Huántar's courtyard.

image of the Chavín deity, which almost resembles the beasts from the children's book *Where the Wild Things Are*. The large figure carved into the rock has big round eyes (looking heavenward), snakes as hair, and a mouth revealing fangs. Its hands show clawlike fingernails. Anthropologists believe fangs could represent the caiman; and the claws, the jaguar, two common animals in the nearby jungle.

The name "El Lanzón" is derived from the Spanish word for "lance," referring to the shape of the stone monolith, which tapers at the bottom. Some scholars argue the shape is actually a fang and not a lance. Lanzón was situated in a central cruciform space underground. Experts have figured out its chief function for the Chavín people: Worshippers, under the influence of a hallucinogenic from the San Pedro cactus (and other psychedelic Amazonian drugs), would be led through the pitch-black underground chambers and hallways. Priests and their assistants would be tooting away at *Strombus galeatus* trumpets, made from conch shells, from hidden spaces in the dark chamber. The thundering sounds of running water from underground streams below would add to the cacophony. After believers roamed the halls in the dark, the climax of this underground journey would be a face-to-face meeting with the deity itself: the priest-guide would lead the pilgrim right up to the monolith, El Lanzón, with its snarled smile and its wide eyes staring deep into the eyes of the probably quivering, superhigh devotee.

Considering that the visitors were under the influence of a hallucinogen, it becomes clear that the temple's details merged into one long, strange trip. For example, stone heads near the entrance were affixed to the temple walls; each one was totally unique, with a combination of human and jaguar features. One of them even had mucus running from his nose, a common side effect for users of various Amazonian hallucinogens.

Having the pilgrims wander through the mazelike underground tunnels while under the influence and hearing a cacophony of noise—is that the sound of the god?— had an immense psychological effect: the staring stone god would frighten visitors while filling them with wonder for this supernatural, spiritual experience.

Some archaeologists believe this ritual was a way to show the ordinary people of this early civilization that there was an order to society, that there was an authority to respect, as evidence has shown that before Chavín's construction, Andean communities lacked these traits. By giving the pilgrims San Pedro cactus, it rendered them vulnerable to the priests, who then could establish a sense of authority, increasing their own political and social power in the society.

Today El Lanzón can once again be visited. Partaking of a cocktail that includes San Pedro cactus may or may not be recommended, though.

Passageways within the main template at Chavín de Huántar.

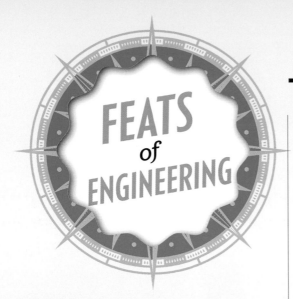

FEATS *of* ENGINEERING

THE GREAT PYRAMID OF GIZA, Egypt

KISH QANAT, Iran

PARIS SEWERS, France

NEW YORK CITY STEAM SYSTEM, USA

THE CHUNNEL, France and England

CERN—EUROPEAN ORGANIZATION FOR NUCLEAR RESEARCH, Geneva, Switzerland

NLESS YOU LIVE ON AN undeveloped island, you're surrounded by so many incredible feats of engineering in the twenty-first-century world that you probably take them for granted. Interstate highways, underground subway systems, skyscrapers that don't just tip over from the sheer weight of glass and concrete. And yet, we humans continue to one-up ourselves with new modern marvels. We've been doing it for thousands of years.

Case in point: the Pyramids of Giza, of course. In 2016, researchers used a new cutting-edge technology in which they could scan the pyramids using infrared thermography, muon radiography imaging, and cosmic particle detectors. They discovered something yet unknown to modern-day humans: two secret chambers buried deep inside the pyramid of Khufu, the largest of the three forty-five-hundred-year-old pyramids located just outside of Cairo.

The two newly discovered chambers are not yet open to the public, but the Great Pyramid does have a fascinating interior that can be visited.

Enter the Great Pyramid about fifty feet up on the north side of the structure. After you traverse a short tunnel, another corridor ascends toward the main chamber. The original contents to the burial chamber are gone—you'll have to go to London, Berlin, and Turin to see them—but there's something invigorating about standing in the belly of the Great Pyramid, no matter how hot and humid the chamber is.

Some media sources, such as History Channel documentaries, are prone to rely on "ancient alien technology" to explain how amazing structures like the pyramids were built. A historian keels over every time that happens. But the Great Pyramid, like other similar structures, was built next to a quarry, making transportation of the heavy stone slightly

A view inside the Grand Gallery within the Great Pyramid of Khufu in Giza, Egypt.

An underground restaurant
on the island of Kish, Iran.

more convenient. Experts believe about two thousand people labored for twenty-three years in constructing this ancient marvel, which stands at nearly five hundred feet tall and contains 2.3 million blocks of limestone. As for how the ancient Egyptians got the precise measurements of the Great Pyramid (and its brethren) so right, they used the position of the stars with the help of a level. No alien help necessary.

THE ENGINEERING TRIUMPH on the Iranian island of Kish, located in the Persian Gulf, is less obvious. Until, of course, you go underground. Kish Qanat (or, as it's sometimes called, Kariz-e-Kish) is a twenty-five-hundred-year-old hydraulic system that drains water from the mountains to the more arid parts of the island, including the ancient city of Harireh. Qanat is a term used in languages throughout the Middle East for "underground canal." Remarkably, Kish is made up entirely of coral, so this tunnel system was carved two and a half thousand years ago right through the coral. The tunnels, fifty-two feet below the surface, snake under the island for five miles.

Though the underground canals were abandoned and replaced by a more modern system, in 1999, developers intent on building an underground city below Kish rediscovered the *qanat* and incorporated the old waterways into the design.

The underground world below Kish today is fascinating, with canals running under arches and alongside pathways, giving the impression that you're strolling through a vast netherworld.

Despite Qanat being so ancient, people don't really go to Kish for its historical ambience. Instead, the island is a major draw for luxury and leisure tourism, its shores crammed with posh resorts.

CENTURIES AFTER KISH Qanat was built, the Paris sewer system was constructed. It's been such an integral part of the city for so long—since the thirteenth century—that it's hard to imagine the City of Light without it. And this is no ordinary sewer system. After all, how many cities do you know that offer tours of the sewer because so many people want to see it?

The first incarnation of city's sewer system was simply a trough that ran down the middle of the street, first constructed around the year 1200. But in 1370, the city built its first underground sewer tunnel below Rue Montmartre. During the reign of Napoleon, there was a massive expansion as the city added 182 more miles of sewer tunnels under the city.

In the early nineteenth century, Pierre Bruneseau, Paris's inspector of works, did something no one had done before: he mapped the Paris's sewer system. And in doing so, he stumbled upon an odd assortment of historical tidbits, including jewelry, gold, and the skeleton of an orangutan that had escaped from the Paris Zoo years earlier. It took Bruneseau seven years to map the system, which, as one would expect, was a dirty job. As the character Jean Valjean says in *Les Misérables*, "Paris has another Paris under herself; a Paris of sewers."

In the middle part of the nineteenth century, there began to be a demand among tourists to see the sewers, and thus started—along with the already strong interest in seeing the city's subterranean catacombs—a fascination with going underground in Paris. In Max Brooks's fantasy novel *World War Z*, Parisians head to the sewers for survival:

Though the thirteen hundred miles of sewage pipe below Paris serve a mundane purpose in residents' everyday life, the sewage system also features heavily in the mystique of the city, as shown by its presence in works such as *Les Misérables* and *World War Z*.

to escape the upcoming zombie apocalypse. Unfortunately, the undead follow them underground.

Today visitors can tour this city under the city. One might not find primate skeletons, but after entering the sewers on the Quai d'Orsay in the Seventh arrondissement, sewer seekers will find the sewer system—be sure to bring some cotton to stick in your nostrils—still running as if it were new. You won't have access to all thirteen hundred miles of sewage pipe—who would want to?—but you'll see enough to make you appreciate this feat of engineering, which was one of the first extensive sewer systems in the world.

An iconic steam vent in New York City.

WHAT THE SEWER system is to Paris, the steam system is to New York City. The system produces steam under the city, which then becomes an agent to cool and heat Gotham's many buildings. When it first began running in 1882, it provided heat and cooling for only lower Manhattan. Today it covers a large swath of Manhattan, particularly in Downtown and Midtown, where the buildings are tallest and can most benefit from the steam system. It is the largest of its kind in the world with 105 miles of pipes and over three thousand manholes. All in all, eighteen hundred buildings in New York City today have their heating and cooling systems powered by steam. Why is the steam system so important? Because it dramatically reduced residents' reliance on coal for heating, thus unclogging the city's air and reducing soot.

Not that it has taken everyone out of danger. Occasionally, part of the steam system explodes. The most recent incident was in 2007, when a steam explosion occurred near Grand Central Terminal, killing one person. For this reason, there are no tours of the underground steam system.

If it weren't for those iconic orange-and-white plastic pipes emitting plumes of steam in the middle of New York City streets—and often seen in movies and TV shows set here—most residents and visitors alike would not realize what's behind their heating and cooling systems.

A CROSS THE ATLANTIC Ocean, one of the world's great engineering miracles occurred on December 1, 1990. That's when Englishman Graham Fagg and Frenchman Philippe Cozette greeted each other underground somewhere below the English Channel. They were helping to excavate what would be known as the Chunnel, the train tunnel linking Britain and France that would make transport between the two countries easier than ever.

Construction of the Chunnel began in 1988, and the first train carrying travelers sped through in 1994. Upon its unveiling that year, the American Society for Civil Engineers named it one of the seven modern wonders of the world. The 31.4-mile-long tunnel is the eleventh

A worker performs maintenance on the Chunnel. The tunnel, connecting England and France, took six years to construct.

longest in the world, with the longest undersea portion of a tunnel on the planet at 23.5 miles.

A tunnel linking France and Britain is not a new idea. It was proposed in the early nineteenth century. The original design included an artificial island about halfway for changing horses. It might have taken them a lot longer than six years to complete the tunnel back then. There are actually three tunnels, or chunnels: two for trains and one for emergency vehicles, all of which are about 160 feet under the seabed and reach 250 feet at their deepest point.

HIDDEN IN THE SWISS Alps is one of the great engineering accomplishments of our lifetime: the accelerator complex at *Conseil Européen pour la Recherche Nucléaire*, better known as CERN. Over 12,000 scientists from seventy different nations work and do research here. The organization is investigating the fundamental laws of nature by colliding particles at the speed of light using the most state-of-the-art scientific tools. Remember Higgs boson, the so-called God Particle that was discovered on July 4, 2012? That was done at CERN. The complex itself is striking: sitting 575 feet below the tip of a mountain on the French-Swiss border, CERN is basically the planet's largest machine. It's made up of a number of accelerators that push particles, injecting said particles to the next accelerator and the next, until finally reaching its final destination: the Large Hadron Collider. The LHC, as it's referred to, is a seventeen-mile-long subterranean tube where particles collide. The collider is made up of almost ten thousand magnets, which create a pull one hundred thousand times stronger than the earth's gravitational pull. And in order to create the proper atmospheric conditions for this process, the inside of the LHC is colder than outer space. The collider tube was approved in 1994 but not finished until 2008, and it was here that the Higgs boson discovery actually took place, a discovery which may—or may not—help us understand the origins of the universe.

Whether or not CERN continues to yield the great answers of the universe, it has already produced life-changing inventions: in 1989, Tim Berners-Lee, working at CERN, invented the Internet as a way to allow scientists to better communicate and share ideas with one another.

All that said, there's a possible dark side to CERN—at least if you conscribe to conspiracy theories. The location of CERN could just be a coincidence, but an intriguing one indeed. The Large Hadron Collider runs underneath the town of Saint-Genis-Pouilly. The word "Pouilly" is derived from the Latin word "Appolliacum," or Apollo. Historians have claimed there was once a Roman temple dedicated to Apollo here. In centuries past, residents believed the temple to Apollo was the entryway to the underworld.

No visitors
beyond this point
Pas de visiteurs
au dela de cette
limite

A look inside the European Organization
for Nuclear Research, better known as
CERN, in Geneva, Switzerland.

INDEX

The Bell Cave in Beit Guvrin-Maresha
National Park, Israel.

DAVID FARLEY IS A NEW YORK–BASED FOOD AND TRAVEL WRITER whose work appears in the *New York Times*, the *Wall Street Journal*, *AFAR*, and the *Guardian*, among other publications. He's the author of the travel memoir *An Irreverent Curiosity: In Search of the Church's Strangest Relic in Italy's Oddest Town*, which was made into a documentary by National Geographic. He has lived in Prague, Rome, and Berlin and has taught writing at New York University.